Tiberius Cavallo

A Complete Treatise on Electricity

Vol. 2

Tiberius Cavallo

A Complete Treatise on Electricity
Vol. 2

ISBN/EAN: 9783337401221

Printed in Europe, USA, Canada, Australia, Japan

Cover: Foto ©berggeist007 / pixelio.de

More available books at **www.hansebooks.com**

. A

COMPLETE TREATISE

ON

ELECTRICITY,

IN

THEORY AND PRACTICE;

WITH

ORIGINAL EXPERIMENTS.

By TIBERIUS CAVALLO, F. R. S.

THE THIRD EDITION,

IN TWO VOLUMES,

Containing the PRACTICE of MEDICAL ELECTRICITY,
befides other ADDITIONS and ALTERATIONS.

VOLUME II.

LONDON:

PRINTED FOR C. DILLY, IN THE POULTRY.

M,DCC.LXXXVI.

A

COMPLETE TREATISE

O N

ELECTRICITY.

P A R T IV.

NEW EXPERIMENTS IN ELECTRICITY.

THE laws of Electricity, together with the experiments neceſſary for their demonſtration, having already been deſcribed, in as compendious a manner as could be done without obſcurity, I ſhall, in the laſt Part of this Work, relate ſuch new experiments and obſervations as I have made at different times, principally with a view to diſcover, if poſſible, the unknown cauſe of ſeveral electrical phenomena, eſpecially thoſe relative to atmoſpherical Electricity.

VOL. II.   B   The

The firſt inſtrument that I made uſe of to obſerve the Electricity of the atmoſphere, was an electrical kite, which I had conſtructed, not with a view to obſerve the Electricity of the air; for this, I thought, was very weak, and ſeldom to be obſerved; but as an inſtrument which could be occaſionally uſed in time of a thunder-ſtorm, in order to obſerve the Electricity of the clouds. The kite, however, being juſt finiſhed, together with its ſtring, which contained a braſs wire through its whole length, I raiſed it on the 31ſt of Auguſt 1775, at ſeven of the clock in the afternoon, the weather being a little cloudy, and the wind juſt ſufficient for the purpoſe. The extremity of the ſtring being inſulated, I applied my fingers to it; which, contrary to my expectation, drew very vivid and pungent ſparks: I charged a coated phial at the ſtring ſeveral times, but I did not then obſerve the quality of the Electricity. This ſuccefsful experiment induced me to raiſe the kite very often, and to keep it up for ſeveral hours together; thinking that if any periodical Electricity, or any change of its quality, took place in the atmoſphere, it might very probably be
diſcovered

difcovered by this inftrument. In the following two Chapters I fhall defcribe the conftruction of the electrical kite, with its appurtenances ; and fhall tranfcribe the moft remarkable part of my journal relative to the kite; *i. e.* defcribing fuch experiments only as are moft remarkable, and do not happen very commonly ; for although I have ufed my kite fometimes ten, and more times in a week, and at any hour of the day or night, yet as the greateft part of thofe experiments are only of ufe to confirm a few laws of atmofpherical Electricity, I fhall omit their particular detail, and fhall only fubjoin thofe laws at the end of the fecond chapter.

CHAP. I.

The Conftruction of the electrical Kite, and other Inftruments ufed with it.

THE firft electrical kite that I conftructed, was feven feet high; and it was made of paper, with a ftick or ftraiter, and a cane-bow, like the kites commonly ufed by fchool-boys. On the upper part of the ftraiter I fixed an iron fpike, projecting about a foot above the kite, which I then thought was abfolutely neceffary to collect the Electricity; and I covered the paper of the kite with turpentine, in order to defend it from the rain. This kite, perfect as I thought it to be in its conftruction, and fit for the experiments for which it was intended, foon manifefted its imperfections, and, after being raifed a few times, it became quite unfit for farther ufe; it being fo large, and confequently heavy, that it could not be ufed, except when the wind was ftrong; and then, after much trouble

in

in raiſing and drawing it in, it often received ſome damage: which ſoon obliged me
to conſtruct other kites upon a different
plan, in order to aſcertain which method
would anſwer the beſt for my purpoſe. I
gradually leſſened their ſize, and varied their
form, till I obſerved, upon trial, that a common ſchool-boy's kite was as good an electrical kite as mine. In conſequence of
which, I conſtructed my kites in the moſt
ſimple manner, and in nothing different
from the children's kites, except that I cc-
vered them with varniſh,, or with well-
boiled linſeed-oil, in order to defend them
from the rain; and I covered the back part
of the ſtraiter with tin-foil, which, however, has not the leaſt power to increaſe its
Electricity. I alſo furniſh the upper extremity of the ſtraiter with a ſlender wire
pointed, which, in time of a thunder-ſtorm,
may perhaps draw the Electricity from the
clouds ſomewhat more effectually; but in
general, I find, as it will appear in the account of the experiments, that it does not
in the leaſt affect the Electricity at the ſtring.
The kites that I generally have uſed are
about four feet high, and little above two

feet

feet wide. This fize, I find, is the moft convenient, becaufe it renders them eafy to be managed, and, at the fame time, they can draw a fufficient quantity of ftring. As for filk or linen kites, they require a good deal of wind to be raifed; and then they are not fo cheap, nor fo eafy to be made, as paper kites are. The ftring fometimes breaks, and the kite is loft, or broken; for which reafon thefe kites fhould be made as cheap and as fimple as poffible.

The ftring is the moft material part of this apparatus; for the Electricity produced is more or lefs, according as the ftring is a better or a worfe Conductor. The ftring which I made for my large kite, confifted of two threads of common twine, twifted together with a brafs wire between the ftrands. This ftring ferved very well for two or three trials; but, on examination, I foon found, that the wire in it was broken in many places, and it was continually fnapping; the metallic continuation, there-fore, being fo often interrupted, the ftring became foon fo bad, that it acted nothing better than common twine without a wire.

I at-

I attempted to mend it, by joining the broken pieces of wire, and working into the twine another wire, which proved a very laborious work; but the remedy had very little effect, the wire breaking again after the first trial; which determined me to adopt other methods : and, after several experiments, I found that the best string was one, which I made by twisting a copper thread * with two very thin threads of twine. Strings like this I have used for the greatest part of my experiments with the kite, and I find them to be exceedingly useful, and fit for the purpose. Silver or gold thread would do much better to twist with the twine, because they are much thinner than copper thread, and, in consequence, the string would be much lighter; but, at the same time, it is to be considered, that gold or silver thread is much dearer than copper thread,

I have attempted to render the twine a good Conductor of Electricity, by cover-

---

* I mean such a thread of copper as is used for trimmings, &c. in imitation of gold threads; which are nothing more than silk, or linen threads, covered with a thin lamina of copper.

ing

ing it with conducting subſtances; as lamp-black, powder of charcoal, very fine emery, and other ſubſtances, mixing them with diluted gum-water; but this method improves the ſtring very little, and for a very ſhort time; for the ſaid conducting ſubſtances are ſoon rubbed off the twine. Mr. Nairne informed me, that he had uſed to ſoak the ſtring of his electrical kite in a ſtrong ſolution of ſalt, which rendered it a good Conductor, ſo far as it attracted the moiſture of the air. In conſequence of this information, I ſoaked in ſalt-water a long piece of twine, and, by raiſing a kite with it, I found that it conducted the Electricity pretty well; but I thought it much inferior to the above-deſcribed ſtring with the copper thread: beſides, the ſalted ſtring in wet weather not only leaves part of the ſalt upon the hands of the operator, and, in conſequence, renders them unfit to manage the reſt of the apparatus, but it marks a white ſpot wherever it touches the clothes.

In raiſing the kite when the weather is very cloudy and rainy, in which time

there

there is fear of meeting with great quantity of Electricity, I generally ufe to hang upon the ftring A B, fig. 9. Plate IV. the hook of a chain C, the other extremity of which falls upon the ground. Sometimes I ufe another caution befides, which is, to ftand upon an infulating ftool; in which fituation I think, that if any great quantity of Electricity, fuddenly difcharged by the clouds, ftrikes the kite, it cannot affect much my perfon. As to infulated reels, and fuch-like inftruments, that fome gentlemen have ufed to raife the kite, without danger of receiving any fhock; fit for the purpofe as they may appear to be in theory, they are yet very inconvenient to be managed. Except the kite be raifed in time of a thunder-ftorm, there is no great danger for the operator to receive any fhock. Although I have raifed my electrical kite hundreds of times without any caution whatever, I have very feldom received a few exceedingly flight fhocks in my arms. In time of a thunder-ftorm, if the kite has not been raifed before, I would not advife a perfon to raife it while the ftormy clouds are juft overhead; the danger in fuch time

being

being very great, even with the precautions
above-mentioned : at that time, without
raifing the kite, the Electricity of the clouds
may be obferved by a cork-ball electrometer
held in the hand in an open place; or, if it
rains, by my electrometer for the rain; `
which will be defcribed hereafter.

When the kite has been raifed, I generally introduce the ftring through a window in a room of the houfe, and faften it
to a ftrong filk lace, the extremity of
which is generally tied to a heavy chair in
the room. In fig. 8. of Plate III. A B reprefents part of the ftring of the kite
which comes within the room; C reprefents the filk lace; D E, a fmall prime
Conductor, which, by means of a fmall
wire, is connected with the ftring of the
kite; and F reprefents the quadrant electrometer, fixed upon a ftand of glafs covered with fealing-wax, which I ufed to
put near the prime Conductor, rather than
to fix it in a hole upon the Conductor, becaufe the ftring A B fometimes fhakes fo
as to pull the prime Conductor down; in
which cafe the quadrant electrometer remains

mains fafe upon the table; otherwife it would be broken, as I have often experienced before I thought of this method. G reprefents a glafs tube, about eighteen inches long, with a knobbed wire cemented to its extremity; with which inftrument I ufe to obferve the quality of the Electricity, when the Electricity of the kite is fo ftrong that I think it not fafe to come very near the ftring. The method is as follows :—I hold the inftrument by that extremity of the glafs tube which is the fartheft from the wire, and touch the ftring of the kite with the knob of its wire, which, being infulated, acquires a fmall quantity of Electricity from it; which is fufficient to afcertain its quality when the knob of the inftrument is brought near an electrified electrometer.

Sometimes, when I raife the kite in the night-time, out of the houfe, and where I have not the convenience of obferving the quality of its Electricity by the attraction and repulfion, or even by the appearance of the electric light, I make ufe of a coated phial, which I can charge at the ftring, and, when charged,

ed, put it into my pocket; wherein it will keep charged even for feveral hours *. By making ufe of this inftrument, I am obliged to keep the kite up no longer than is necef-fary to charge the phial, in order to obferve the quality of the Electricity in the atmo-fphere; for after the kite has been drawn in

* The conftruction of this phial is as follows :— Befides the coating of the infide and outfide, that this phial has, like any other of the fame kind, a glafs tube, open at both ends, is cemented into its neck, and pro-ceeds within the phial, having a fmall wire faftened to its lower extremity, which touches the infide non-elec-tric coating. The wire with the knob of this phial is cemented into another glafs tube, which is nearly twice as long, and fmaller than the tube cemented into the neck of the phial. The wire is cemented fo, that only its knob projects out of one end, and a fmall length of it out of the other end of the tube. If this piece with the wire be held by the middle of the glafs tube, it may be put in or out of the tube which is in the neck of the phi-al, fo as to touch the fmall wire at the lower extremity of it, and that without difcharging the phial, if it is charged. I have kept fuch a phial charged for fix weeks together; and probably it would keep much longer, if it were to be tried. The ingenious young Electrician may make ufe of fuch a phial for feveral diverting pur-pofes.—The piece of glafs which ferves to hold the wire by, is rather better to be fixed above than below the ball. In this cafe, the ball is perforated quite through, and the wire projects a fhort way above it; to which the glafs tube is cemented.

8

and]

and brought home, I can then examine the
Electricity of the infide of the phial, which
is the fame as that of the kite.

When the Electricity of the kite is very
ftrong, I fix a chain, communicating with
the ground, at about fix inches diftance
from the ftring; which may carry off its
Electricity, in cafe that this fhould increafe
fo much as to put the by-ftanders in danger.

Befides the above-defcribed apparatus, I
have occafionally ufed fome other inftruments, which I have often varied, according as fome particular experiments required; but, as they are of no great confequence, I fhall omit to defcribe them. It
is only neceffary, before I enter into the
narration of the principal experiments
performed with the kite, to give an idea of
the ftandard of my quadrant electrometer;
which may, very probably, fhew the fame
intenfity of Electricity under a number of
degrees different from the other inftrument
of the fame kind. When the kite is flying,
and the apparatus is difpofed as in fig. 8. of
Plate

Plate III, I bring, under the extremity E of the prime Conductor, a little bran, held upon a tin plate, and obferve, that when the index of the electrometer is at ten degrees, the prime Conductor begins to attract the bran at the diftance of about three-fifths of an inch; when the index is at twenty degrees, the prime Conductor attracts the bran at the diftance of about one inch and a quarter; when the index is at thirty degrees, the bran begins to be attracted at the diftance of two inches and one-fifth. Thefe diftances vary, as the weather changes its degree of drynefs; but in frofty weather I obferve them conftantly as above.

CHAP.

C H A P. II.

Experiments performed with the electrical Kite.

SEptember the 2d, 1775. The weather being very cloudy, and actually raining, the kite was raifed at eight o'clock P. M. with two hundred yards of ftring, which had a brafs wire through its whole length. The wind was from the fouth and very ftrong. The Electricity at the ftring was negative, and juft fufficient to charge a half-pint phial fo as to give a fhock fenfible to the elbows. The kite, after being up for about one hour, fell to the ground, having its paper, which was not properly varnifhed, almoft entirely torn off by the violence of the wind and the rain.

September the 14th. The kite was raifed with a ftrong north wind at half paft three P. M. The Electricity was pofitive, and

and pretty ftrong, the index of the electro-meter being generally about 20°*. The weather was rather cold, and very thick clouds were gradually approaching the ze-nith. The kite was pulled down at half paft four P. M.

N. B. At night the aurora borealis was very ftrong, and feveral flafhes of lightning were feen near the horizon towards the north.

September the 23d. A fmall kite was raifed at half paft ten o'clock in the morn-ing, and it was kept up for eleven hours fucceffively, *viz.* till half paft nine P. M. The ftring, which was only a common twine, without a wire, was conftantly elec-trified pofitively, although in a very fmall degree. About nine o'clock the Electri-city appeared ftronger, fo that a fmall phi-al, charged at the ftring, gave a pretty fen-fible fhock. The weather was very clear, and warm; but in the night no aurora bo-

* The index of the electrometer in general rifes higher or falls lower, according as the kite comes nearer to, or goes farther from the zenith; the length of the ftring remaining the fame.

realis,

realis, or any other electrical appearance, was perceived. The wind was east by south, and so weak that the kite was kept up with great difficulty.

October the 10th, 1775. The weather being clear, and the wind blowing strong from the south west, the kite was raised at eleven o'clock A. M. with ninety yards of string, which had a copper thread twisted in *. The wind, during the experiment, increased and decreased several times; and the Electricity, which was positive, as it appeared by the index of the electrometer, also increased and decreased. At noon the violence of the wind caused the kite to fall. At half past four o'clock, the wind being a little more moderate, the kite was raised again. The Electricity was also positive, and seemed rather stronger than it had been in the morning. The weather at this time was cloudy; the clouds appearing much thicker near the horizon than about the zenith. The kite was pulled down at half

* Such string as this was used in all the following experiments.

paſt five o'clock, and at half after ſeven
was raiſed again; every phenomenon continuing the ſame. At eight o'clock, while
I was pulling the kite in, I inſulated the
ſtring when only thirty-five yards of it
were out, and was ſurpriſed to find, that
now the Electricity was as ſtrong as it had
been, when all the ſtring was out, which
was ninety yards long. It muſt however
be remarked, that at this time a few flaſhes
of lightning were ſeen among the clouds,
which were pretty thick about the horizon.
At a quarter paſt eleven o'clock the kite
was raiſed again, which was the fourth
time of raiſing it that day; the weather
then being very clear, and the wind the
ſame as in the afternoon. The Electricity
was very weak, but conſtantly poſitive.
The kite was pulled down, after having
been up a few minutes only.

October the 16th. At about two P. M.
a thick fog being juſt cleared up, the weather became clear, and the wind began to
blow from the ſouth-ſouth-weſt. The kite
was raiſed with one hundred and twenty
yards of ſtring, and it was kept up no
longer

longer than a quarter of an hour. The Electricity was pofitive, and pretty ftrong; the index of the electrometer being about 15°. At half paft three o'clock the kite was raifed again, the weather being very little cloudy. At half paft four o'clock the clouds became very thick, and in a fhort time it began to rain, which increafed the Electricity of the kite, without changing its quality; the index of the electrometer arriving to 20°. The kite was pulled down at five o'clock.

October the 18th. After having rained a great deal in the morning, and night before, the weather became a little clear in the afternoon, the clouds appearing feparated, and pretty well defined. The wind was weft, and rather ftrong, and the atmofphere in a temperate degree of heat. In thefe circumftances, at three P. M. I raifed my electrical kite with three hundred and fixty feet of ftring. After that the end of the ftring had been infulated, and a leather-ball, covered with tin-foil, had been hung to it, I tried the power and quality of the Electricity, which ap-

peared

peared to be pofitive, and pretty ftrong.   In
a fhort time a fmall cloud paffing over, the
Electricity increafed a little ; but the cloud
being gone, it decreafed again to its former ,
degree.   The ftring of the kite was now
faftened by the filk lace to a poft in the'
yard of the houfe wherein I lived, which
was fituated near Iflington, and I was repeatedly charging two coated phials, and
giving fhocks with them :—while I was fo
doing, the Electricity, which was ftill pofitive, began to decreafe, and in two or three
minutes time it became fo weak, that it
could be hardly perceived with a very fenfible cork-ball electrometer.   Obferving at
the fame time that a large and black cloud'
was approaching the zenith (which, no
doubt, caufed the decreafe of the Electricity) indicating imminent rain, I introduced the end of the ftring through a window,
in a firft-floor room, wherein I faftened
it by the filk lace to an old chair.   The
quadrant electrometer was fet upon the
fame window, and was, by means of a wire,
connected with the ftring of the kite.   Being now three quarters of an hour after
three o'clock, the Electricity was abfolutely

lutely unperceivable ; however, in about
three minutes time, it became again perceivable, but now upon trial was found to
be negative; it is therefore plain, that its
ſtopping was nothing more than a change
from poſitive to negative, which was evidently occaſioned by the approach of the
cloud, part of which by this time had
reached the zenith of the kite, and the rain
alſo had begun to fall in large drops.—The
cloud came farther on ;—the rain increaſed, and the Electricity keeping pace with
it, the electrometer ſoon arrived to 15°.
Seeing now, that the Electricity was pretty
ſtrong, I began again to charge the two
coated phials, and to give ſhocks with
them ; but the phials had not been charged
above three or four times, before I perceived that the index of the electrometer
was arrived to 35°, and was keeping ſtill
increaſing. The ſhocks now being very
ſmart, I deſiſted from charging the phials
any longer ; and, conſidering the rapid advance of the Electricity, thought to take off
the inſulation of the ſtring, in caſe that if
it ſhould increaſe farther, it might be ſilently conducted to the earth, without cauſing

C 3   any

any bad accident, by being accumulated in the infulated ftring. To effect this, as I had no proper apparatus near me, I thought to remove the filk lace, and faften the ftring itfelf to the chair ; accordingly, I difengaged the wire that connected the elec-trometer with the ftring ; laid hold of the ftring; untied it from the filk lace, and faftened it to the chair ; but while I effected this, which took up lefs than half a minute of time, I received about a dozen, or fifteen, very ftrong fhocks, which I felt all along my arms, in my breaft, and legs; fhaking me in fuch a manner, that I had hardly power enough to effect my purpofe, and to warn the people in the room to keep their diftance. As foon as I took my hands off the ftring, the Electricity (in confequence of the chair being a bad Conductor) began to fnap between the ftring and the fhutter of the window, which was the neareft body to it. The fnappings, which were audible at a good diftance out of the room, feemed firft ifochronus with the fhocks which I had received, but in about a minute's time, oftner ; fo that the people of the houfe compared their found to the rattling noife of a

jack

jack going when the fly is off. The cloud now was juſt over the kite; it was black, and well defined, of almoſt a circular form, its diameter appearing to be about 40°; the rain was copious, but not remarkably heavy. As the cloud was going off, the electrical ſnapping began to weaken, and in a ſhort time became unaudible. I went then near the ſtring, and finding the Electricity weak, but ſtill negative, I inſulated it again, thinking to keep the kite up ſome time longer; but obſerving that another larger and denſer cloud was approaching apace towards the zenith, as I had then no proper apparatus at hand, to prevent every poſſible bad accident, I reſolved to pull the kite in; accordingly a gentleman, who was by me, began pulling it in, while I was winding up the ſtring. The cloud was now very nearly over the kite, and the gentleman, who was pulling in the ſtring, told me, that he had received one or two ſlight ſhocks in his arms, and that if he were to feel one more, he would certainly let the ſtring go; upon which I laid hold of the ſtring, and pulled the kite in as faſt as I

could,

could, without any farther obfervation; being then ten minutes after four o'clock.

N. B. There was neither thunder or lightning perceived that day, nor indeed for fome days before or afterwards.

November the 8th, 1775. The wind being north-weft, and juft fufficient, the kite was raifed at three quarters paft eleven A. M. with one hundred and twenty yards of ftring. The Electricity was pofitive, and weak, the weather being cloudy. At noon the clouds grew thicker, and the Electricity quite vanifhed; however, in a few feconds it returned, and from this time it evidently kept increafing and de-creafing, according as the clouds became thinner or thicker. At forty minutes after one o'clock the Electricity vanifhed again; a thick cloud then covering almoft the whole hemifphere; but as a little rain be-gan to fall, the Electricity returned, and it was ftill pofitive. At three quarters paft three o'clock the clouds began to grow thin, and the Electricity increafed a little; but at this time I was obliged to pull the
kite

kite in. The index of the electrometer in this experiment seldom arrived to 6°.

November the 16th. The weather being very clear and frosty, the kite was raised at a quarter past ten A. M. with one hundred and twenty yards of string. The Electricity was positive, and pretty strong, the index of the electrometer going from 9° to 15°; raising as the wind blew stronger, and the kite was more elevated, and *vice versa*. At a quarter past three o'clock the wind, which was north-north-west, intirely failing, the kite fell.

November the 17th. The weather being exceedingly damp, and the fog so dense, that the houses at about a quarter of a mile distance could not be distinguished, the kite was raised at two P. M. with one hundred and ten yards of string, while it was raining, but very little. The Electricity was positive, and so weak that the cork-balls of an electrometer diverged about three quarters of an inch. The wind being very violent, I was obliged to pull the kite in, after having been up for about five minutes.

December

December the 5th, 1775. The weather being equally cloudy, and the wind weſt by north, and hardly ſufficient, the kite was raiſed at a quarter paſt three P. M. with one hundred and twenty yards of ſtring. The Electricity was poſitive, and ſo weak as to cauſe the cork-balls of an electrometer to diverge about an inch. At a little after four o'clock the kite was pulled in; and at eight o'clock in the evening it was raiſed again. At this time the Electricity was much ſtronger than in the afternoon, but conſtantly poſitive. The weather clearing up, the clouds were driven away by the wind, which was now a little ſtronger than in the afternoon. At forty minutes after eight o'clock the ſky was clear, the moon and ſtars appearing very bright; except that a few thin clouds were yet to be ſeen near the horizon. The index of the electrometer was now going from 15 to 20°. At ten minutes after nine o'clock the kite was drawn in.

N. B. No aurora borealis was to be ſeen.

December

December the 20th. The weather being cloudy and hazy, the kite was raifed at
three quarters after ten o'clock A. M. with
~~one~~ hundred and forty yards of ftring.
The Electricity was pofitive, and pretty
ftrong, the index of the electrometer going
from 16° to 21°. At half paft one P. M.
the weather growing a little clearer, I pulled the kite down ; and, after having interpofed a filk ribband between its loop and
the extremity of the ftring, fo as to infulate the kite, I raifed it again with the fame
length of ftring; and, after I had infulated
the lower extremity of the ftring, I obferved that the intenfity of the Electricity,
as it appeared by the index of the electrometer, was, as nearly as could be determined, the fame as before, *i. e.* when
the kite was not infulated with refpect to
the ftring.

At two o'clock P. M. I pulled the kite
down, and found, upon obfervation, that the
filk ribband had contracted no moifture, fo
that the kite was perfectly infulated by it.
This experiment of infulating the kite I
have often repeated at other times, and

have

have always met with the fame fuccefs; hence it appears, that it is the ftring, and not the kite, which in general collects the Electricity from the air. The kite there- fore in general is only ufeful to extend the ftring high into the open air.

January the 4th, 1776. The froft hav- ing been very hard during the day and night before, the wind began to blow very ftrong from the fouth at two o'clock A. M. which occafioned a fudden thaw and a copious rain. At eight o'clock A. M. at which time the kite was raifed, the hemifphere appeared like a uniform dark canopy, un- der which feveral fmall, irregular, and darker clouds were running very faft; the rain was conftant, but not remarkably heavy. As foon as the ftring of the kite was infulated, the Electricity, which was negative, began to fnap from it, to the fhutter of the window and other bodies near; the index of the electrometer arrived to 40°, and it would have certainly gone farther, if the apparatus had been drier; but the air was fo damp, that it was almoft impoffible to keep any part of the appara-

tus

tus fufficiently free from moifture. The Electricity, however, gradually decreafed, fo that at ten o'clock A. M. at which time the kite was pulled in, the index of the electrometer was at a little above 12°. The coated phials in this experiment were charged furprifingly quick; three or four feconds of time being fufficient to charge two half-pint phials completely.

January the 11th. The ground was covered with ice and fnow, and the atmo-fphere was fo hazy, that the houfes at a mile diftance could not be perceived. The wind was fouth-eaft by fouth, and juft fuf-ficient to raife the kite, which was raifed at three o'clock P. M. with one hundred and twenty-four yards of ftring, and kept up till half an hour after midnight. When the kite was firft raifed it began to thaw, but as foon as it was dark it began to freeze again very hard. The Electricity was po-fitive, and pretty ftrong, the index of the electrometer being about 13°. At half paft four o'clock I let out thirty-four yards more of ftring, fo that all the ftring the kite now had, was one hundred and fifty-eight yards.

yards. With this addition of ftring the Electricity increafed, fo that the index of the electrometer arrived to 17°. At half after five o'clock the wind began to increafe, and the Electricity to decreafe, until the index of the electrometer arrived to 6°. At three quarters paft fix o'clock the index of the electrometer was about 13°, and at feven o'clock it arrived to 20°; the wind .being now quite eaft. At a quarter paft feven o'clock the index of the electrometer was about 25°. From this time the wind and the Electricity began both to decreafe, fo that at nine o'clock the index of the electrometer was about 10°. At eleven o'clock the wind increafed. At twelve o'clock the wind was very ftrong, and the index of the electrometer was about 6°. At half paft twelve o'clock the index of the electrometer was between 3° and 4°; but the wind being grown very violent, the ftring broke very near the window, and was loft with the kite.

N. B. A few minutes after the kite was loft, it began to fnow copioufly.

January

January the 26th.   The froſt being very intenſe, as it had been for about three weeks, and actually ſnowing, I raiſed the kite, with ſeventy yards of ſtring; but before the ſtring was inſulated, it ceaſed to ſnow, and the weather began to clear up, and ſoon became very ſerene.   The Electricity was poſitive, and very ſtrong, the index of the electrometer being about 32°. At eleven o'clock the ſtring broke, and the kite fell, after having been up for above three quarters of an hour.

February the 17th, 1776.   The weather being cloudy, rainy, and ſo hazy, that the houſes at half a mile diſtance could not be diſcerned, the kite was raiſed at three quarters paſt eleven o'clock A. M. with one hundred and ſeventy-five yards of ſtring.   The wind was pretty ſtrong; the Electricity was negative, and alſo ſtrong; the index of the electrometer being about 20°.   In about five minutes time the rain ceaſed, the wind weakened, and ſhifted a little towards the ſouth; and the Electricity changed from negative to poſitive. The index of the electrometer was now about

about 15°. In two or three minutes time, it began to rain again, and continued fo for the greateft part of that day; the wind became very weak, and the Electricity changed again from pofitive to negative, and continued fo till half an hour after noon; at which time the wind became fo weak that I was obliged to pull the kite in.

February the 19th. The fky being full of pretty well defined clouds, and the wind weft-north-weft, the kite was raifed at half paft three o'clock P. M. with one hundred and feventy-five yards of ftring. The Electricity was pofitive and ftrong, the index of the electrometer going from 10° to 20°. At three quarters paft three o'clock a denfe cloud paffed over the kite, which occafioned the index of the electrometer to defcend to 4°. As the cloud went away, the electrometer elevated its index. At four o'clock the kite was pulled down.

April the 8th, 1776. The weather was clear, and the northern light very ftrong.
The

The kite was raifed for a few minutes at nine o'clock P. M. with one hundred and feventy-five yards of ftring, the wind being north-north-weſt, and pretty ftrong. The Electricity was pofitive, and, as I could judge, the index of the electrometer would have arrived to 15°.

May the 15th, 1776. The weather being cloudy, and the wind north, the kite was raifed at three o'clock P. M. with one hundred and feventy yards of ftring. The Electricity was at firft exceedingly weak, and, as I imagine (for I had not time to examine it) pofitive. But a denfe cloud paffing over the kite, the Electricity vanifhed; and, as a few drops of rain fell, a very weak negative Electricity appeared, which foon increafed, fo as to caufe the index of the electrometer to arrive to 15°. The rain, however, in a few minutes ceafed, and the Electricity gradually decreafed and vanifhed. A very weak pofitive Electricity immediately took place; but, as another denfer cloud paffed over, and a few very fmall drops of rain fell, the pofitive Electricity vanifhed, and the ne-

gative

gative took place. The cloud and rain foon went off, and the Electricity became again pofitive, and continued fo till the kite was pulled down. According as the clouds, which paffed continually over the kite, were thinner or thicker, fo the Electricity was more or lefs intenfe; fometimes caufing the index of the electrometer to arrive to 5°, and at other times being fcarce perceivable with the cork-ball electrometer. At five o'clock the kite was pulled in, the weather being then pretty clear, and the index of the electrometer at 3°. The wind, during this experiment, was ftronger or weaker, according as the clouds which paffed over were thicker or thinner. At half paft feven o'clock in the evening of the fame day, the kite was raifed again, with the fame length of ftring, the wind being then rather ftrong, and the weather pretty clear. The Electricity was pofitive, and the index of the electrometer ftood at 10°; but as fome clouds came from the north, the Electricity began to decreafe, and by eight o'clock it juft feparated the balls of an electrometer, the hemifphere being then entirely covered by clouds. At half

4

paft

paſt eight o'clock the kite was pulled down, the clouds over the kite being then very thin, and the index of, the electrometer at 5°.

June the 4th, 1776. The weather being cloudy, and the wind on the ſouth-ſouth-weſt, the kite was raiſed at one o'clock P. M. with one hundred and ſeventy yards of ſtring. The Electricity was poſitive, and the index went from 1° to 7°. At three quarters paſt one o'clock the clouds began to be diſſipated, and the Electricity increaſed a little. At two o'clock the kite was pulled in.

June the 17th. The weather being cloudy, and the wind ſouth-weſt, the kite was raiſed at five o'clock P. M. with one hundred and ſeventy yards of ſtring. The Electricity was poſitive, and the index of the electrometer went from 10° to 16°. In this experiment the clouds, whether thicker or thinner, ſeemed to have no ef-fect upon the Electricity of the kite. At a quarter paſt ſix o'clock the kite was pulled in.

D 2   June

June the 20th. The weather being
cloudy, and the wind eaft, and juft fufficient, the kite was raifed at three quarters paft three P. M. with one hundred
and feventy yards of ftring. The Electricity was pofitive, and the index of the
electrometer ftood about 8°. At five
o'clock the weather began to clear up, and
the Electricity to increafe; fo that in half
an hour's time, the index of the electrometer arrived to 17°; and at fix o'clock it
ftood at 25°. But the wind fuddenly failing about this time, the kite fell.

January the 8th, 1777. The weather
being frofty and clear, and the wind north,
and pretty ftrong, the kite was raifed at
four o'clock P. M. with one hundred and
feventy yards of ftring. The Electricity
was pofitive and ftrong, the index of the
electrometer being at 36°. The fpark taken
from the fmall prime Conductor was remarkably pungent in this experiment, although it was hardly a quarter of an inch
long. At a quarter paft five o'clock the
kite was pulled in.

General

General Laws, deduced from the Experiments
performed with the electrical Kites.

I. The air appears to be electrified at all
times ; its Electricity is constantly positive,
and much stronger in frosty, than in warm
weather * ; but it is by no means less in
the night than in the day-time †.

II. The presence of the clouds generally
lessens the Electricity of the kite : sometimes it has no effect upon it ; and it is
very seldom that it increases it a little,

* My observations upon the Electricity of the atmosphere have been made in almost every degree of
temperature, from 15° to 80° of FARENHEIT's ther-
mometer.

† In all my experiments, it happened only once
that the string of the kite gave no signs of Electricity ;
it was one afternoon, when the weather was warm, and
the wind so weak, that the kite was raised with difficulty, and could hardly be kept up for a few minutes ;
in the evening, however, the wind, which in the day-
time had been north-west, shifted to the north-east,
blowing a little stronger ; I then raised the kite again,
being half past ten o'clock, and obtained, as usual, a
pretty strong positive Electricity.

D 3   III. When

III. When it rains, the Electricity of the kite is generally negative, and very seldom positive.

IV. The aurora borealis seems not to affect the Electricity of the kite.

V. The electrical spark taken from the string of the kite, or from any infulated Conductor connected with it, especially when it does not rain, is very seldom longer than a quarter of an inch; but it is exceedingly pungent. When the index of the electrometer is not higher than 20°, the person that takes the spark will feel the effect of it in his legs; it appearing more like the discharge of an electric jar, than the spark taken from the prime Conductor of an electrical machine.

VI. The Electricity of the kite is in general stronger or weaker, according as the string is longer or shorter; but it does not keep any exact proportion to it: the Electricity, for instance, brought down by a string of a hundred yards, may raise the index of the electrometer to 20°; when,

with

with double that length of ftring, the index of the electrometer will not go higher
than 25°.

VII. When the weather is damp, and
the Electricity is pretty ftrong, the index
of the electrometer, after taking a fpark
from the ftring, or prefenting the knob of
a coated phial to it, rifes furprifingly quick
to its ufual place; but in dry and warm
weather, it rifes exceeding flow.

Thefe few laws are, in fhort, the deduction of all my experiments performed with
the kites, during the courfe of about two
years. How far they may be of ufe, or
may coincide with the obfervations of other
experimentators, I will not pretend to fay.
My experiments have been performed at
Iflington, and perhaps the refult of fimilar
ones may be different at other places, efpecially under different climates; I wifh,
therefore, that they may be accurately repeated in other places, and their refult
may be compared together; in order to determine, if poffible, fomething fatisfactory,
relative to the caufe of that perpetual

Electricity

Electricity which exifts in the atmofphere, and which, very probably, occafions the Electricity of the clouds.

CHAP. III.

Experiments performed with the Atmofpheri-
cal Electrometer, and the Electrometer for
the Rain.

FIG. 1. of Plate III, reprefents a very fimple inftrument, which I have contrived for the purpofe of making obfervations on the Electricity of the atmofphere; and which, on feveral accounts, feems to be the moft ufeful for that purpofe. A B is a common jointed fifhing-rod, without the laft or fmalleft joint. From the extremity of this rod proceeds a flender glafs tube C, covered with fealing-wax, and having a cork D, at its end, from which a pith-ball electrometer is fufpended. H G I is a piece of twine faftened to the other extremity of the rod, and fupported at G by a fmall ftring F G.

At

At the end I of the twine a pin is faftened, which, when pufhed into the cork D, renders the electrometer E uninfulated.

When I would obferve the Electricity of the atmofphere with this inftrument, I thruft the pin I into the cork D, and holding the rod by its lower end A, pro-ject it out from a window in the upper part of the houfe into the air, raifing the end of the rod with the electrometer, fo as to make an angle of about 50° or 60° with the horizon.   In this fituation I keep the inftrument for a few feconds, and then pulling the twine at H, I difengage the pin from the cork D ; which operation caufes the ftring to drop in the dotted fituation K L, and leaves the electrometer infulated and electrified, with an Electricity contrary to that of the atmofphere.—This done, I draw the inftrument into the room, and examine the quality of the Electricity, without obftruction either from wind or darknefs.

With this inftrument I have made ob-fervations on the Electricity of the atmo-fphere,

fphere, feveral times in a day for feveral months, and from them I have deduced the following general obfervations, which feem to coincide with thofe made with the electrical kites.

I. That there is in the atmofphere, at all times, a quantity of Electricity; for, whenever I ufe the above-defcribed inftrument, it always acquires fome Electricity.

II. That the Electricity of the atmofphere, or fogs, is always of the fame kind, namely pofitive; for the electrometer is always negative, except when it is evidently influenced by heavy clouds near the zenith; as appears by the obfervations made the 19th of October, in the following fpecimen of the journal.

III. That in general, the ftrongeft Electricity is obfervable in thick fogs, and alfo in frofty weather; and the weakeft, when it is cloudy, warm, and very near raining: but it does not feem to be lefs by night than in the day time.

IV. That

IV. That in a more elevated place the Electricity is ſtronger than in a lower one ; for, having tried the atmoſpherical electrometer, both in the ſtone and iron gallery in the cupola of St. Paul's Cathedral, I found that the balls diverged much more in the latter than in the former leſs elevated place ; hence it appears, that, if this rule takes place at any diſtance from the earth, the Electricity in the upper regions of the atmoſphere muſt be exceedingly ſtrong.

The

The following is the most remarkable part of the journal of the observations made with the above-described atmospherical electrometer, in which I have noted the Electricity of the atmosphere, *i. e.* the contrary of that in the atmosphere.

N. B. The stroke ——— signifies *as above*.

Time of observation.	Clouds.	Fog.	Wind.	Opening of the electrometer in inches.	Quality of the Electricity.
October the 19th, 1776, ½ past 10 o'clock	Cloudy	Very little at a distance	Very strong	$\frac{1}{10}$	Negative
11	Heavy clouds	———	Violent	$\frac{1}{4}$	Positive
¼ past 2	Less clouds	———	Little	1	Negative
3 ¼ past	Few at a distance	———		$\frac{1}{2}$	
November the 6th, 11 o'clock, P. M.	o	Exceedingly thick	o	1	
November the 8th, 12 at noon		Hardly any		$\frac{1}{4}$	
11		Very little			
November the 13th, 10 o'clock, A. M.	Cloudy	Little at a distance	Pretty strong	Very little	
10 P. M.	o	o	Violent	$\frac{1}{4}$	
November the 17th, 11 o'clock, P. M.			Very little	$\frac{1}{4}$	
12			Little		
November the 28th, 10 o'clock, A. M.	Cloudy	Little	Hardly any	$\frac{1}{4}$	
2		o	Very little		
December the 20th, ½ past 9 o'clock, P. M.	———	Little	Very little	———	
February the 6th, 1777, 2 o'clock, P. M.	o	Very little	Hardly any	$\frac{1}{4}$	
February the 7th, 12 at noon	Few on the north	o	Hardly any	Very little	
8	o		o		
February the 27th, 12 o'clock, P. M.				$\frac{1}{10}$	
March the 26th, 11 o'clock, P. M.					

The electrometer for the rain, in principle is nothing more than an insulated instrument to catch the rain, and by a pithball electrometer to shew the quantity and quality of its Electricity.

Fig. 2. of Plate III. represents an instrument of this kind, which I have frequently used, and after several observations, have found to answer very well. A B C I is a strong glass tube about two feet and a half long, having a tin funnel, D E, cemented to its extremity, which funnel defends part of the tube from the rain. The outside surface of the tube from A to B is covered with sealing-wax; so also is the part of it which is covered by the funnel. F D is a piece of cane, round which several brass wires are twisted in different directions, so as to catch the rain easily, and at the same time, to make no resistance to the wind. This piece of cane is fixed into the tube, and a slender wire proceeding from it goes through the bore of the tube, and communicates with the strong wire A G, which is thrust into a piece of cork fastened to the end A of the tube. The end G of the wire A G is formed into a ring, from which

I suspend

I fufpend a more or lefs fenfible pith-ball electrometer, as occafion requires.

This inftrument is faftened to the fide of the window-frame, where it is fupported by ftrong brafs hooks at C B, which part of the tube is covered with a filk lace, in order to adapt it better to the hooks. The part F C is out of the window, with the end F a little elevated above the horizon. The remaining part of the inftrument comes through a hole in one of the lights of the fafh, within the room, and no more of it touches the fide of the window than the part C B.

When it rains, efpecially in paffing fhowers, this inftrument, ftanding in the fituation above defcribed, is frequently electrified ; and, by the diverging of the electrometer, the quantity and quality of the Electricity of the rain may be obferved, without any danger of a miftake. With this inftrument I have obferved, that the rain is generally, though not always, elec- trified negatively, and fometimes fo ftrong- ly that I have been able to charge a fmall coated phial at the wire A G.

This

This inftrument fhould be fixed in fuch a manner, that it may be eafily taken off from the window, and replaced again, as occafion requires; for it will be neceffary to clean it very often, particularly when a fhower of rain is approaching.

I fhall conclude this chapter with the defcription of a pocket electrometer, fig. 5 and 6. of Plate III. that I have lately conftructed, and which, on feveral accounts, feems preferable to thofe of the moft fenfible fort now in ufe. The cafe, or handle of this electrometer is formed by a glafs tube, about three inches long, and three-tenths of an inch in diameter, half of which is covered with fealing-wax. From one extremity of this tube, *i. e.* that without fealing-wax, a fmall loop of filk proceeds, which ferves occafionally to hang the electrometer on a pin, &c. To the other extremity of the tube a cork is adapted, which, being cut tapering on both ends, can fit the mouth of the tube with either end. From one extremity of this cork, two linen threads proceed, a little fhorter than the length of the tube, fufpending each a little *cone* of pith of elder. When this

this electrometer is to be uſed, that end of the cork which is oppoſite to the threads is puſhed into the mouth of the tube; then the tube forms the inſulated handle of the pith electrometer, as repreſented fig. 6. Plate III. · But when the electrometer is to be carried in the pocket, then the threads are put into the tube, and the cork ſtops it, as repreſented fig. 5. The peculiar advantages of this electrometer are, its convenient ſmall ſize, its great ſenſibility, and its continuing longer in good order than any other I have yet ſeen.

Fig. 4. of Plate III. repreſents a caſe to carry the above-deſcribed electrometer in. This caſe is like a common tooth-pick caſe, except that it has a piece of amber fixed on one extremity A, which may occaſionally ſerve to electrify the electrometer negatively, and on the other extremity it has a piece of ivory faſtened upon a piece of amber BC. This amber BC ſerves only to inſulate the ivory, which, when inſulated, and rubbed againſt woollen cloths, acquires a poſitive Electricity; and it is therefore uſeful to electrify the electrometer poſitively.

CHAP.

Experiments made with the Electrophorus,
commonly called a Machine for exhibiting
perpetual Electricity.

IN fig. 9. of Plate III. there are represented some Plates, commonly called
the Machine for exhibiting perpetual Electricity, or the *Electrophorus*. This machine
consists of two plates, one of which, B, is
a circular glass plate, covered on one side
with some sulphureous or resinous electric,
most commonly with a composition made
of equal parts of rosin, shell-lac, and sul-
phur; the other plate A, is a brass plate,
or a board covered with tin-foil, which is
nearly of the same dimensions as the electric plate, and it is furnished with a glass
handle I, which, by means of a brass
or wooden socket, is screwed into its
center. This machine is the invention of
an Italian philosopher (Mr. VOLTA of
Como) and its use is the following :

First, the plate B is excited, by rubbing its coated side with a piece of new white flannel, and when excited as much as poffible, is fet upon the table with the coated side uppermoft. Secondly, the metal plate is laid upon the excited electric, as reprefented in the figure. Thirdly, the metal plate is touched with the finger or any other Conductor, which, on touching the plate, receives a fpark from it. Laftly, the metal plate A, being held by the extremity of its glafs handle I, is feparated from the electric plate; and, after it is elevated above that plate, it will be found ftrongly electrified, with an Electricity contrary to that of the electric plate, in which cafe it will give a very ftrong fpark to any Conductor brought near it. By fetting the metal upon the electric plate, touching it with the finger, and feparating it fucceffively, a great number of fparks may be obtained apparently of the fame ftrength, and that without exciting again the electric plate. If thefe fparks are repeatedly given to the knob of a coated phial, this will prefently become charged.

The

The action of thefe plates depends upon a principle long ago difcovered, *viz*. the power that an excited electric has to induce a contrary Electricity in a body brought within its fphere of action ; the metal plate, therefore, when fet upon the excited electric, acquires a contrary Electricity, by giving its electric fluid to the hand, or other Conductor that touches it, when fet upon a plate pofitively electrified ; or acquiring an additional quantity of fluid from the hand, &c. when fet upon a plate electrified negatively.

As to the continuance of the virtue of this electric plate, when once excited, without repeating the excitation, I think there is not the leaft foundation for believing it perpetual, as fome perfons have fuppofed ; it being nothing more than an excited electric, it muft gradually lofe its power, by imparting continually fome of its Electricity to the air, or other fubftances contiguous to it. Indeed its Electricity, although it could never be proved to be perpetual by experiments, lafts a very long time, it having been obferved to be pretty ftrong

E 2   feveral

feveral days, and even weeks after excitation. The great duration of the Electricity of this plate, I think, depends upon two caufes: firft, becaufe it does not lofe any Electricity by the operation of putting the metal plate upon it, &c.; and fecondly, becaufe of its flat figure, which expofes it to a lefs quantity of air, in comparifon with a ftick of fealing-wax, or the like, which being cylindrical, expofes its furface to a greater quantity of air, which is continually robbing the excited electrics of their virtue.

The firft experiments that I made relative to this machine, were with a view to difcover which fubftance would anfwer beft for coating the glafs plate, in order to produce the greateft effect. I tried feveral fubftances either fimple or mixed, and at laft I obferved, that the ftrongeft in power, as well as the eafieft I could conftruct, were thofe made with the fecond fort of fealing-wax *, fpread upon a thick plate of glafs.

* It is remarkable, that fometimes they will not act well at firft; but they may be rendered very good, by fcraping with the edge of a knife the fhining or glof-
fy

glafs †. A plate that I made, after this manner, and no more than fix inches in diameter, when once excited, could charge a coated phial feveral times fucceffively, fo ftrong as to pierce a hole through a card with the difcharge. Sometimes the metal plate, when feparated from it, was fo ftrongly electrified, that it darted ftrong flafhes to the table upon which the electric plate was laid, and even into the air, befides caufing the fenfation of the fpider's web upon the face brought near it, like an electric ftrongly excited. The power of fome of my plates is fo ftrong, that fometimes the electric plate adheres to the metal, when this is lifted up; nor will they feparate, even if the metal plate is

fy furface of the wax. This feems analogous to the well-known property of glafs ; which is, that new cylinders or globes, made for electrical purpofes, are often very bad electrics at firft, but that they improve by being worked, *i. e.* by having their furface a little worn. Paper alfo has nearly the fame property.

† I have lately feen fome of thofe plates, conftructed by Mr. G. ADAMS, which acted exceedingly well ; and they were made with a compofition of two parts of fhell-lac, and one part of Venice turpentine, without any glafs plate.

E 3

touched

touched with the finger, or other Conductor.

If, after having excited the fealing-wax, I lay the plate with the wax upon the table, and the glafs uppermoft, *i. e.* contrary to the common method, then, on making the ufual experiment of putting the metal plate on it, and taking the fpark, &c. I cbferve it to be attended with the contrary Electricity; that is, if I lay the metal plate upon the electric one, and while in that fituation touch it with an infulated body, that body acquires the pofitive Electricity, and the metallic, removed from the electric plate, appears to be negative; whereas it would become pofitive if laid upon the excited wax. This experiment, I find, anfwers in the fame manner, if an electric plate is ufed which has the fealing-wax coating on both fides, or one of Mr. Adams's, which has no glafs plate.

If the brafs plate, after being feparated from, be prefented with the edge toward the wax, lightly touching it, and thus be drawn over its furface, I find that the

Electricity

Electricity of the metal is abforbed by the fealing-wax, and thus the electric plate lofes part of its power; and if this operation is repeated five or fix times, the electric plate lofes its power intirely, fo that a new excitation is neceffary in order to revive it.

If, inftead of laying the electric plate upon the table, it is placed upon an electric ftand, fo as to be accurately infulated, then the metal plate fet on it acquires fo little Electricity that it can only be difcovered with an electrometer; which fhews that the Electricity of this plate will not be confpicuous on one fide of it, if the oppofite fide is not at liberty either to part with, or acquire more of the electric fluid. In confequence of this experiment, and in order to afcertain, how the oppofite fides of the electric plate would be affected in different circumftances, I made the following experiments :

Upon an electric ftand E, fig. 9. Plate III. I placed a circular tin plate, nearly fix inches in diameter, which by a flender

wire

wire H communicated with an electrometer of pith-balls G, which was also infulated upon the electric ftand F. I then placed the excited electric plate D, of fix inches and a quarter in diameter, upon the tin plate, with the wax uppermoft, and on removing my hand from it, the electrometer G, which communicated with the tin plate, *i. e.* with the under fide of the electric plate, immediately opened with negative Electricity. If, by touching the electrometer, I took that Electricity off, the electrometer did not afterwards diverge. But if now, or when the electrometer diverged, I prefented my hand open, or any other uninfulated Conductor, at the diftance of about one or two inches, over the electric plate, without touching it, then the pith-balls diverged; or if they diverged before, came together, and immediately diverged again with pofitive Electricity;—I removed the hand, and the balls came together;—approached the hand, and they diverged; and fo on.

If, while the pith-balls diverged with negative Electricity, I laid the metal plate,

holding

holding it by the extremity K of its glafs handle, upon the wax, the balls came, for a little time, towards one another, but foon opened again with the fame, *i. e.* negative Electricity.

If, whilft the metallic refted upon the electric plate, I touched the former, the electrometer immediately diverged with pofitive Electricity, which if, by touching it, I took off, the electrometer continued without divergence. — I touched the metal plate again, and the electrometer opened again ; and fo on for a confiderable number of times, until the metal plate had acquired its full charge. On taking now the metal plate up, the electrometer G inftantly diverged with ftrong negative Electricity.

I repeated the above-defcribed experiments with this only difference in the difpofition of the apparatus, *i. e.* I laid the electric plate D, with the excited fealingwax, upon the circular tin plate, and the glafs uppermoft ; and the difference in their refult was, that where the Electricity had been

been pofitive in the former difpofition of the apparatus, it now became negative, and *vice verfa*; except that, when I firft laid the electric plate upon the tin, the electrometer G diverged with negative Electricity, as well in this as in the other difpofition of the apparatus.

I repeated all the above-mentioned experiments with an electric plate, which, befides the fealing-wax coating on one fide, had a ftrong coat of varnifh on the other fide; and their refult was fimilar to that of thofe made with the above-defcribed plate.

As to the explanation of thefe experiments, they feem to depend upon thefe two well-known principles, *viz.* that a body brought within the fphere of action of an electrified body, does actually acquire the contrary Electricity : and that the exiftence of one kind of Electricity upon the furface of a fubftance whatever, caufes the exiftence of the contrary Electricity upon fome other fubftance near it.

CHAP. V.

Experiments on Colours.

HAVING accidentally obferved that an electric fhock, fent over the fur-face of a card, marked a black ftroke upon a red fpot of the card, I was from this in-duced to try what would be the effect of fending fhocks over cards painted with dif-ferent water-colours; accordingly I paint-ed feveral cards with almoft every colour I had, and fent fhocks * over them, when they were very dry; making ufe of the uni-verfal difcharger, fig. 5. Plate I. The ef-fects were as follow:

Vermilion was marked with a ftrong black track, about one-tenth of an inch wide. This ftroke is generally fingle, as reprefented by A B, fig. 7. of Plate III.; fometimes it is divided in two towards the

* The force generally employed was the full charge of one foot and a half of coated glafs.

middle,

middle, like E F ; and fometimes, particularly when the wires are fet very diftant from one another, the ftroke is not continued, but interrupted in the middle, like G H. It often, although not always, happens, that the impreffion is marked ftronger at the extremity of that wire from which the electric fluid iffues, as it appears at E, fuppofing that the wire C communicates with the pofitive fide of the jar; whereas the extremity of the ftroke, contiguous to the point of the wire D, is neither fo ftrongly marked, nor furrounds the wire fo much as the other extremity E.

Carmine received a faint and flender impreffion of a purple colour.

Verdigrife was fhook off from the furface of the card, except when it had been mixed with ftrong gum-water, in which cafe it received a very faint impreffion.

White lead was marked with a ftrong black track, not fo broad as that on vermilion.

Red

Red lead was marked with a faint mark, much like carmine.

The other colours I tried, were orpiment, gambodge, fap-green, red-ink, ultramarine, Pruffian blue, and a few others, which were compounds of the above; but they received no impreffion.

It having been infinuated, that the ftrong black mark, which vermilion receives from the electric fhock, might poffibly be owing to the great quantity of fulphur contained in that mineral, I was induced to make the following experiment :—I mixed together equal quantities of orpiment and flower of fulphur, and with this mixture, by the help, as ufual, of very diluted gum-water, I painted a card; but the electric fhock fent over it left not the leaft impreffion.

Defirous of carrying this inveftigation on colours a little further, with a particular view to determine fomething relative to the properties of lamp-black and oil *, I procured

* It has often been obferved, that when the lightning has ftruck the mafts of fhips, it has paffed over
                                                        fuch

cured fome pieces of paper painted on both fides with oil colours, and fending the charge of two feet of coated glafs over each of them, by making the interruption of the circuit upon their furfaces, I obferved that the pieces of paper painted with lamp-black, Pruffian-blue, vermilion, and purple brown, were torn by the explofion, but white lead, Naples yellow, Englifh ochre, and verdigrife remained unhurt.

The fame fhock fent over a piece of paper painted very thickly with lamp-black and oil left not the leaft impreffion. I fent the fhock alfo over a piece of paper unequally painted with purple brown, and the paper was torn where the paint laid very thin, but remained unhurt where the paint was evidently thicker. Thefe experiments I repeated feveral times, and with fome little variation, which naturally produced

fuch parts of the mafts which were covered with lamp-black and tar, or painted with lamp-black and oil, without the leaft injury, at the fame time that it has fhivered the uncoated parts, in fuch a manner as to render the mafts ufelefs. For a particular account of fuch facts, fee the Phil. Tranf. vol. XLVIII. and LXVII.

different

different effects; however, they all feem to point out the following propofition.

I. A coat of oil paint, over any fubftance, defends it from the effects of fuch an electric fhock, as would otherwife injure it; but by no means defends it from any electric fhock whatever. II. No one colour feems preferable to the others, if they are equal in fubftance, and equally well mixed with oil; but a thick coating does certainly afford a better defence than a thinner one.

By rubbing the above-mentioned pieces of paper, I find that the paper painted with *lamp-black and oil* is more eafily excited, and acquires a ftronger Electricity, than the papers painted with the other colours; and perhaps on this account it may be, that lamp-black and oil might refift the fhock fomewhat better than the other paints.

It is remarkable, that vermilion receives the black impreffion, when painted with linfeed-oil, nearly as well as when painted with water. The paper painted with white lead and oil, receives alfo a
black

black mark; but its nature is very fingular. The track, when firft made, is almoft as dark as that marked on white lead, painted with water, but it gradually lofes its blacknefs, and in about one hour's time (or longer, if the paint is not frefh) it appears without any darknefs; and when the painted paper is laid in a proper light, appears only marked with a colourlefs track, as if made by a finger-nail. I fent the fhock alfo over a piece of board, which had been painted with white lead and oil about four years before, and the explofion marked the black track upon this alfo; this track however was not fo ftrong, nor vanifhed fo foon as that marked upon the painted paper, but in about two days time it alfo vanifhed intirely.

CHAP.

C H A P. VI.

Promiſcuous Experiments.

OBſerving that a ſtrong ſpark may be obtained from the metal plate belonging to Mr. VOLTA's machine, deſcribed in the fourth chapter of this Part, when not the leaſt ſpark can be obtained from the electric plate itſelf, I was naturally induced to make uſe of the above-mentioned metallic plate, to diſcover the Electricity of very weak electrics; which otherwiſe would be either inobſervable, or ſo ſmall as not to permit its quality to be aſcertained. Accordingly I conſtructed ſeveral ſuch plates of different ſizes, beginning from that of a common metal button faſtened upon a ſtick of ſealing-wax; and by uſing them, I obtained a very ſenſible Electricity from the hairs of my legs, when ſtroked, and of my head, or any part that I have tried of my body, or the head of almoſt any other perſon.

VOL. II.   F   In

In this manner I obtain fuch ftrong fparks from the back of a cat, a hare's fkin, a rabbit's fkin, a piece of flannel, or of paper, that I can prefently charge a coated phial with either of thofe, and fo ftrongly, as to pierce a hole through a card with its difcharge.

I have often obferved that, when ftroking a cat with one hand, and holding it with the other, I feel frequent fmart pricklings on different parts of that hand, which holds the animal. In thefe circumftances very pungent fparks may be drawn from the tips of the ears of the cat.

Smooth glafs rubbed with a rabbit's fkin, dry and warm, acquires, I find, the *negative* Electricity ; but if the fkin is cold, the glafs is excited pofitively. Sometimes fmooth glafs may be excited negatively with new white flannel, clean and dry, and alfo with a hare's fkin.

Obferving the ftrong electric power of new white flannel, I thought that a piece of it, rolled round the globe of an electrical machine,

machine, would perhaps give a ftronger Electricity to the prime Conductor than the glafs itfelf. In order to try the truth of my fuppofition, I tied a large piece of flannel, dry and warm, round the globe of the machine, and for a rubber, I applied the palm of my hand; then turned the winch, firft flowly, and afterwards brifkly; but, contrary to my expectation, the Electricity at the prime Conductor, although pofitive, was fo weak, that the index of the quadrant electrometer was not moved from its perpendicular fituation. Surprifed at this event, I refolved to take off the apparatus; but I was more furprifed, when, on removing the flannel from the globe, the former appeared fo ftrongly pofitive, that it darted feveral fparks to my arm, and other contiguous bodies, and the latter remained fo ftrongly negative, that the electrometer upon the prime Conductor inftantly elevated its index to about 45°. This experiment being feveral times repeated, produced always the fame effect.

Having had occafion to coat a ten ounce phial for the Leyden experiment, I ftuck

the

the brafs filings on the infide of it with varnifh, agreeable to the directions given by fome writers on Electricity. - This phial remained about a week unufed, but it happened that whilft I was charging and difcharging it for fome experiments, on making a difcharge, it exploded with a greater noife than ufual, the cork with the wire being at the fame time blown out of the neck of it. Being intent upon the main experiments in hand, I omitted to examine this phenomenon ; — I replaced the cork into the neck of the phial, and went on charging and difcharging it again; but it had not been charged above three or four times more, when, on making a difcharge, the varnifh that ftuck the brafs filings was in a flame, which burnt the under fide of the cork, and occafioned a good deal of fmoke and flame to come out of the phial. Some days after, this experiment was repeated in the prefence of three gentlemen, well verfed in Electricity, when the cork with the wire was alfo pufhed out of the neck of the phial; but the varnifh was this laft time fo far burnt, that the brafs filings were almoft all dropped to

the

the bottom of the phial, and had their colour changed by the combuftion.

In making fome experiments, of a nature
rather different from Electricity, I accidentally obferved, that when I agitated fome
quickfilver in a glafs tube, hermetically
fealed, and in whofe cavity the air was very
much rarefied, the outfide of the tube appeared fenfibly electrified; its Electricity
however was not conftant, nor, as I firft
thought, in proportion to the agitation of
the quickfilver. Being defirous of afcertaining the properties of fuch tubes, I conftructed feveral of them, and by means of
two cork-ball electrometers, obferved their
properties; but as they all agree in regard
to the chief points, I fhall only defcribe
one, which is the beft of them. This
tube is reprefented by fig. 3. of Plate III.
Its length is thirty-one inches, and its diameter is little lefs than half an inch. The
quickfilver in it may be about three
fourths of an ounce; and in order to exhauft it of air, I clofed it while the quickfilver was boiling in its oppofite end.

Before

. Before this tube is ufed, I make it a little warm, and clean it; then holding it nearly horizontal, I let the quickfilver in it run from one end of the tube to the other, by gently and alternately elevating and deprefling its extremities.. This operation immediately renders the outfide of the tube electrical, but with the following remarkable property, *viz.* that end of the tube, where the quickfilver actually ftands, is pofitive, and all the remaining part is negative. If by elevating this pofitive end of the tube a-little, I let the quickfilver run to the oppofite end, which was negative, then the former inftantly becomes nega-tive, and the latter pofitive. The pofitive end has always a ftronger Electricity than the negative. If when one end of the tube, for inftance A, is pofitive, *i. e.* when the quickfilver is in it, I do not take off that Electricity by touching it; then on elevating this end A, fo as to let the quickfilver run to the oppofite end B, it appears negatively electrified in a very fmall degree. If by depreffing it again it be rendered pofitive a fecond time, and that pofitive Electricity is neither taken off, then on elevat-

ing

ing this end A again, it appears to be pofitive in a fmall degree : but if whilft it is
pofitive, its Electricity be taken off, then
on being elevated, it appears ftrongly negative.

When about two inches of each extremity of this tube is coated with tin-foil as
it appears in the figure, that coating affifts
to render the Electricities at the extremities
of the tube more confpicuous, fo that fometimes they give fparks to a Conductor
brought near.

In regard to the conftruction of fuch
tubes (which I have made of feveral lengths,
from nine to thirty-one inches) it is obfervable, that fome will act very well,
while others will hardly acquire any Electricity at all, even when they are made
very hot. I am not yet thoroughly fatis-
fied in refpect to this difference, but fufpect that the thicknefs of the glafs is more
concerned, than any thing elfe; it appearing that a tube, whofe glafs is about one-
twentieth of an inch thick, anfwers better than either a thicker or a thinner one.

F 4   The

The following Chapters contain the account of fome experiments, which I made fince the firft publication of this Book; for which reafon, I deemed proper to relate them apart, without altering the preceding chapters, with the contents of which they are however connected.

C H A P. VII.

An Account of fome new Experiments in Electricity, with the Defcription and Ufe of two new Electrical Inftruments.

PRofeffor LICHTENBERG of Gottingen, fome time ago made an experiment upon the electrophorus, an account of which was firft received in London towards the latter end of the year 1777. The phenomena attending the experiment are very entertaining and various, but I do not know that any perfon ever offered a fatisfactory explanation of them. The author himfelf, in his paper entitled " *De nova methodo naturam ac motum fluidi electrici invefligandi Commentatio prior,*" wherein he

gives

gives an account of the experiment, does
not attempt any explanation of it ; content-
ing himfelf with the account only of various particulars attending it.—In brief, the
experiment is as follows : ·

The electrophorus, that is, a plate of
fome refinous fubftance, as fulphur, rofin,
gum-lac, &c. is firft excited, either by
rubbing or otherwife; then a piece of metal of any fhape, at pleafure, as for inftance, a three-legged compafs, a piece of
brafs tube, or the like, is fet upon the electrophorus, and to this piece of metal fo
placed, a fpark is given, of the Electricity
contrary to that of the plate; this done,
the piece of metal is removed, by means
of a ftick of fealing-wax or other electric,
and fome powder of rofin, kept in a linen
bag, is fhaken upon the electrophorus : this
powder will be found to fall about thofe
points upon the plate, which the piece of
metal touched, forming fome radiated appearances, much like the common repre-
fentations of ftars; at the fame time, upon
the greateft part of the plate, that is, befides thofe ftars, there is hardly any powder

at .

at all. Now, it is to be remarked, that if the plate is excited negatively, and the fpark given to the metal fet upon it is pofitive, the appearance will be as above-defcribed ; but if, on the contrary, the plate is pofitive and the fpark is negative, then the powder of rofin will be found to fall upon thofe parts of the plate which in the other cafe it left un-covered, and to leave the ftars clean ; in fhort, it will do juft the reverfe of what it did in the other cafe ; or, in other words, the powder of rofin will be attracted by thofe parts only of the electrophorus which are electrified pofitively,

When I firft obferved thefe phenomena, I thought there was no apparent reafon why the powder of rofin fhould be attracted by thofe parts of the electrophorus which are pofitively electrified, and not by thofe which are negative. The two Electricities are certainly contrary to one another ; but either of them attracts a non-electrified body. Infifting upon this confideration, I thought that the experiment could be explained only upon the fuppofition, that the powder of rofin, on its falling from the

linen

linen bag was actually electrified negatively; in which cafe it would have been eafy
to account for the phenomena upon the
well-known principle of bodies contrarily
electrified attracting each other, and repelling one another when poffeffed of the
fame kind of Electricity.

In order to try the reality of my fuppofition by experiments, I infulated a brafs
plate upon a glafs ftand, and connected a
very fenfible electrometer with it; then
began fhaking the powder of rofin upon it,
in the fame manner as I had done upon the
electrophorus, and in a few feconds time
had the pleafure to fee the electrometer diverge with a very manifeft degree of negative Electricity, anfwering my expectations
exactly. The explanation of the ingenious
Profeffor LICHTENBERG's experiment,
now became very eafy and natural; for the
powder of rofin being actually electrified
negatively, could not be attracted, except by thofe parts of the electrophorus
which are in a contrary ftate, *i. e.* electrified pofitively. It is obferved, that powder
of rofin anfwers better for this experiment
than

than the powders of other fubftances; and accordingly I find that this powder, when fhaken upon the infulated brafs plate, fhews a ftronger degree of Electricity than the other powders. Indeed the Electricity of the powder of rofin, not only when fhaken upon the brafs plate in the manner above-mentioned, but fimply let fall upon it from a piece of paper, a fpoon, &c. is very great; half an ounce of this powder being fuffici-ent, to let the threads of the electrometer diverge as much as they poffibly can.

This difcovery not only affords an eafy explanation of Profeffor Lichtenberg's experiment upon the electrophorus, but fhews a method of exciting powders, which has long been a defideratum in the fcience of Electricity. The method is as follows : —Infulate a metal plate upon an electric ftand, and connect with it a cork-ball electrometer; then the powder required to be tried, being held in a fpoon, or other thing, at about fix inches above the plate, is to be let fall gradually upon it. In this manner the Electricity acquired by the pow-der, being communicated to the metal plate,

plate, and to the electrometer, is rendered manifeſt by the divergence of the threads; and its quality may be aſcertained in the uſual manner. See fig. the 4th of Plate IV.

It muſt be obſerved, that if the powder is of a conducting nature, like the amalgam of metals, ſand, &c. it muſt be held in ſome electric ſubſtance, as a glaſs phial, a plate of ſealing-wax, or the like. Sometimes the ſpoon that holds the powder may be inſulated; in which caſe, after the experiment, the ſpoon will be found poſſeſſed of an Electricity contrary to that of the powder.

In performing theſe experiments care muſt be had to render the powders, and whatever they are held in, as free from moiſture as poſſible; ſometimes it being neceſſary to make them very warm, otherwiſe the experiment is apt to fail. The following are the particulars that I have obſerved with this new method, which however are neither numerous, nor often repeated; but they may ſuffice to excite
the

the curiosity of thofe perfons, who have leifure and the opportunity of repeating them more at large and in a greater variety.

Powder of rofin, whether it be let fall from paper, glafs, or a metal fpoon, electrifies the plate ftrongly negative ; the fpoon, if infulated, remaining ftrongly pofitive. Flower of fulphur produces the fame effect, but in a little lefs degree. Pounded glafs, let fall from a piece of paper, made dry and warm, electrifies the plate negatively, but not in fo ftrong a degree as rofin. If it is let fall from a brafs cup, it electrifies the plate pofitively, but in a very fmall degree.

Steel-filings let fall either from a glafs phial or paper, electrify the plate negatively ; but brafs-filings, treated in the fame manner, electrify the plate pofitively. The amalgam of tin-foil and mercury, gunpowder, or very fine emery, electrify the plate negatively, when they are let fall upon it from a glafs phial. Quickfilver, from a glafs a glafs phial, electrifies the plate pofitively.

Soot from the chimney, or the afhes of common pit-coals mixed with fmall cinders, electrify the plate negatively, when let fall from a piece of paper.

Defcription of the improved atmofpherical Electrometer.

Fig. the 2d of Plate IV. is a geometrical representation of my new atmofpherical electrometer, in its real fize; this inftrument, whofe firft hint I received from my friend T. RONAYNE, Efq; after various trials, I brought to the prefent ftate of perfection, as long ago as the year 1777; and immediately after, feveral of them were made after my pattern by Mr. ADAMS, philofophical inftrument maker, in Fleet Street. The great difficulty attending the conftruction of this inftrument, has long diffuaded my publifhing any defcription of it; nor had I ever prefented the defcription of it to the Royal Society, if the obfervations of feveral of my friends, who have ufed it, in England and abroad, joined to my own repeated experiments, had not indifputably confirm-

7   ed

ed its fuperiority over any other inftrument of that kind.   Its particular advantages are : I. The fmallnefs of the fize * ; II. Its being always ready for experiments, without fear of entangling the threads, or having an equivocal refult by the fluggifhnefs of its motion ; III. Its being not difturbed by wind or rain ; IV. Its great fenfibility ; and V. Its keeping the communicated Electricity longer than any other Electrometer.

The principal part of this inftrument is a glafs tube C D M N, cemented at the bottom into the wooden piece A B, by which part the inftrument is to be held when ufed for the atmofphere ; and it alfo ferves to fcrew the inftrument into its wooden cafe A B O, fig. I. when it is not to be ufed +.   The upper part of the tube C D M N, is fhaped tapering to a fmaller extremity, which is entirely covered with fealing-wax, melted by heat, and not dif-

* Sometime ago I made one fo fmall, that its cafe which is of brafs, meafures only 3 ¼ inches in length and $\frac{9}{10}$ of an inch in diameter, and yet it acts exceeding well.

+ The whole cafe of this electrometer has been, made alfo of brafs, and has been found to anfwer better than wood, as it does not warp.

8

folved

solved by spirits. Into this tapering part
a small tube is cemented, the lower extremity G of which being also covered with
sealing-wax, projects a short way within
the tube C D M N. Into this smaller
tube a wire is cemented, which with its
lower extremity touches the flat piece of
ivory H, fastened to the tube by means of
cork : the upper extremity of the wire
projects about a quarter of an inch above
the tube, and screws into the brass cap
E F, which cap is open at the bottom, and
serves to defend the waxed part of the instrument from the rain, &c. In fig. 3.
a section of this brass cap is represented,
in order to shew its internal shape, and the
manner in which it is screwed to the wire
projecting above the tube L. The small
tube L, and the upper extremity of the
large tube C D M N, appear like one continued piece, on account of the sealing-
wax, which covers them both. The conical corks P of this electrometer, which
by their repulsion shew the Electricity, &c.
are as small as they can possibly be made,
and they are suspended by exceedingly fine

filver wires ; thefe wires are fhaped in
a ring at the top, by which they hang
very loofely to the flat piece of ivory H,
which has two holes for that purpofe.
By this method of fufpenfion, which is
applicable to every fort of electrometer,
the friction is leffened almoft to nothing,
and thence the inftrument is fenfible of a
very fmall degree of Electricity.   I M; and
K N, are two narrow flips of tin-foil,
ftuck to the infide of the glafs C D M N,
and communicating with the wooden bottom A B ;—they ferve to convey off that
Electricity, which, when the corks touch
the glafs, is communicated to it, and being
accumulated, might difturb the free motion
of the corks.

In regard to its ufe, this inftrument may
ferve to obferve the artificial, as well as the
atmofpherical Electricity.   When it is to
be ufed for artificial Electricity, this electrometer is fet upon a table or other convenient fupport; then it is electrified by
touching the brafs cap E F with an electrified body, which Electricity will fometimes

times be preferved for more than an hour *;
in this ftate, if any electrified fubftance is
brought near the cap E F, the corks of the
electrometer, by their converging or diverging more, will fhew the fpecies of that
body's Electricity.

Before we proceed, it is neceffary to remark, that to communicate any Electricity
to this electrometer, by means of an excited electric, *e. g.* a piece of fealing-wax
(which we fuppofe as always negatively
electrified) is not very readily done in the
ufual manner, becaufe of the cap E F being
well rounded, and free from points or fharp
edges. By the approach of the wax, the
electrometer will be caufed to diverge; but
as foon as the wax is removed, the wires
immediately collapfe. The beft method to
electrify it, is to bring the excited wax fo
near the cap, that one or both the corks
may touch the fide of the bottle C D M N;
after which, they will foon collapfe and appear unelectrified; if now the wax is re-

* I once made an electrometer of this fort, which
could continue to be electrified for above 12 hours,
and that in a room without fire.

moved,

moved, they will again diverge, and remain electrified positively.

In this operation, the wax does not impart any of its Electricity to the electrometer, but only acts by means of its atmosphere, *viz*. when the excited wax is first brought near the brass cap E F, (agreeable to the well-known* law of Electricity, and according to Dr. Franklin's hypothesis) it determines the electric fluid naturally belonging to the corks, towards the cap; hence the corks repel each other. Now, if in this state they touch the sides of the glass C D M N, they acquire from it a quantity of electric fluid equal to that which, by the action of the excited wax, was driven towards the cap; consequently they collapse, and appear unelectrified. Notwithstanding this appearance, the cap is actually overcharged; so that when the wax is removed, the overplus of the electric fluid, which the corks had acquired from the glass and tin-foil stuck upon it, and which was crowded upon the cap, because of the negative atmosphere of the wax, now diffuses itself equally through the cap,

6

the

the wires, the corks, &c. and therefore the corks repel each other with pofitive Electricity.

If, inftead of the fealing-wax excited negatively, an electric poffeffed of pofitive Electricity be ufed, the electrometer acquires the negative Electricity, and the explanation, *mutatis mutandis*, is the fame as above.

By confidering this remark it will appear, that when this electrometer is electrified either pofitively or negatively, and an electrified body be brought towards the brafs cap, the Electricity of that body will be of the fame kind with that of the electrometer if the corks increafe their divergency ; but it will be of the contrary kind if the corks approach one another.

When this inftrument is to be ufed to try the Electricity of the fogs, the air, the clouds, &c. the obferver is to do nothing more than to unfcrew it from its cafe, and, holding it by the bottom A B, to prefent it to the open air, a little above his head, fo

G 3　　　　　　that

that he may conveniently fee the corks P, which will immediately diverge if there is any fufficient quantity of Electricity; whofe nature, *i. e.* whether pofitive or negative, may be afcertained by bringing an excited piece of fealing-wax, or other electric, towards the brafs cap E F.

It is perhaps unneceffary to remark, that this obfervation muft be made in an open place, as the roads out of town, the fields, the top of a houfe, &c.

In the roads between Iflington and London, I have often made ufe of this inftrument: by which I have confirmed the obfervations of Thomas Ronayne, Efq; who firft difcovered the Electricity of the fogs, as teftified by a paper of his publifhed in the Phil. Tranfactions; and who has remarked, that a fog is very rarely not electrified, but in frofty weather the air is conftantly electrified.

Promifcuous Experiments.

Having had frequent occafion to obferve how difficult it is to deprive fealing-wax

4

of

of its Electricity entirely, after it has been
well excited, I had the curiofity to try
whether water could effect it; in order to
that, I tied a ftick of fealing-wax to a
filk ftring, about a yard long, and after
having excited it very powerfully with flannel, I plunged it in a tin veffel full of water, and immediately drawing it out,
brought a very fenfible electrometer near it,
and obferved, that at firft it fhewed no fign
of Electricity, but in about half a minute's
time it manifefted a fmall, but very fenfible
degree of negative Electricity. A glafs
tube treated in the fame manner, was deprived of all its Electricity by the water.

I have often remarked, that after having
excited a glafs tube with the amalgamed
rubber, in the ufual manner, the part of it
which had been under my hand was negative. This minus ftate was ftill more confpicuous, when I grafped with my hand
the part next above, *viz.* part of that
which had been excited pofitively by rubbing. In the fame manner, when I excite
a ftick of fealing-wax by rubbing it with
flannel, I often find, that the part of it
G 4   which

which I had held with my hand was in a contrary ftate of Electricity, *viz.* pofitive.

Being defirous of trying the conducting power of the effluvia of burning bodies, in a manner more fatisfactory than it had hitherto been done, I contrived an inftrument for that purpofe, which is reprefented in fig. 5 *. The handle of it, A B, is a glafs tube, into the extremity B of which, a wire E I, and a fmaller glafs tube B C, are cemented by means of fealing-wax. From the extremity of this fmall tube, another wire G F proceeds, which, as well as the wire E I, is bent at top, fo that the extremities of both wires E F may be about one-tenth of an inch from one another. G H is a fmall wire, faftened to the wire F G, and to the extremity of the handle, fo that when the inftrument is held in one's hand, this wire touches the hand. K is a fmall cork-ball electrometer, which, when the inftrument is to be ufed, is affixed to the pin D, which proceeds from the wire I E. When ex-

* This fig. is half the real fize of the inftrument.

periments

periments are to be tried with this inftrument, the electrometer K muft be affixed to
the pin D, and muft be electrified fo as the
cork-balls may diverge as far as poffible :
this done, the extremities E F of the wires
are brought within the effluvia that are to
be tried, which, if they are of a good conducting nature, will complete the communication between the two wires E F, and
difcharge the electrometer of its Electricity ; otherwife the electrometer will remain electrified for a confiderable time.
The experiments which I made with this
inftrument are neither numerous, nor fo
often repeated as to be depended upon ;
excepting one only, which perhaps it will
not be ufelefs to mention : I found that
the fumes arifing by the action of a lens
from the amalgam of tin-foil and mercury,
conducted fo badly, that the electrometer
loft its Electricity in a time very little lefs
than is required without any fumes whatever.

C H A P.

CHAP. VIII.

Experiments concerning the Effects of Electri-
city in Vacuo.

BEFORE we begin with the narration
of the experiments made with a view
of afcertaining the effects of Electricity in
vacuo, it will be proper to mention the ftate
of knowledge relating to electric attraction
and repulfion, conducting power, and the
appearance of electric light in vacuo.

The inquifitive Mr. BOYLE, towards the
latter end of the laft century, obferved, that
excited electrics would attract in the vacuum
of his air-pump; in confequence of which
he concluded that the prefence or abfence
of air did not interfere with electric attraction. In the beginning of the prefent century Mr. GREY repeated Mr. BOYLE's experiments, and, like him, found that electrics would attract at nearly the fame diftance in vacuo as in air. He likewife
made fome other experiments in vacuo,
concerning

concerning electric attraction and repulsion, from which the same deductions could be inferred. After Mr. GREY divers other ingenious persons repeated such-like experiments, and came to very nearly the same conclusion; but F. BECCARIA seems to have been the first person who asserted, that in a perfect vacuum there is no electric attraction, and his assertion is certainly true.

As for the electric light in vacuo, numerous observations concerning its diffusibility and various shades of its colours in a moderate degree of exhaustion have been made with sufficient accuracy by Mr. HAUKSBEE, Mr. du FAY, Abbé NOLLET, F. BECCARIA, and others, who likewise observed that the vacuum was a conductor of electricity; but it is related by Dr. PRIESTLEY, in his first volume of experiments and observations on different sorts of permanently electric fluids, that Mr. WALSH, assisted by Mr. de LUC, having made a double barometer, in which the quicksilver had been accurately boiled so as to expel all the air from the tube, found that the vacuum in the arched part of this double barometer

was

was not a conductor of Electricity, nor any
electric light could be seen in it. This
remarkable difcovery was lately confirmed
by fome ingenious experiments of Mr.
Morgan, defcribed in the 75th vol. of the
Phil. Tranf.; but my experiments, which are
related in this chapter, were made not with
a torricellian vacuum, but with an excellent
air-pump, *viz.* that recommended in the
preceding part of this work *.

Experiment I.

In a glafs receiver, of fix inches diameter
and nine inches in height, having a brafs
cap, a brafs wire of $\frac{2}{10}$ of an inch in diameter
was fixed to its cap, and proceeding through
the middle of the receiver, its lower extremity was five inches diftant from the aperture of the receiver, and of courfe of the
plate of the air-pump, when the receiver
was placed upon it. A fine linen thread
was faftened towards the top of the wire, and
about four inches of it hanged freely along
the brafs wire, and almoft in contact with it.

* An account of thofe experiments was read at the
Royal Society in November 1784.

The

The extremity of the wire, which paffing
through the brafs cap projected out of the
receiver, was furnifhed with a ball.   Thus
prepared, the receiver was placed upon the
plate of the pump, without any leather, or
any thing elfe befides a little oil on its outfide edge, which muft be always underftood
in all the other experiments related in the
courfe of this chapter.   Then the exhauftion was commenced, and at intervals fome
Electricity was communicated, either by the
approach of the Conductor of an electrical machine, or the knob of a charged jar, to the brafs
ball of the wire, in order to obferve the ftrength
of the repulfion of the thread from the wire in
different degrees of rarefaction, which degrees were afcertained by the fhort baro-
metrical gage.   Proceeding in this manner,
it was obferved, that till the rarefaction did
not exceed one hundred, to wit, till the
air remaining within the receiver was not
léfs than the hundredth part of the original
quantity, whenever the Electricity was communicated to the brafs ball, the thread firft
adhered to the wire, and then was repelled
by it; though this repulfion became fmaller
and fmaller, according as the exhauftion
came

came nearer to the above-mentioned degree.
The clinging of the thread to the wire firft,
was becaufe, being dry, it required fome time
before it acquired a fufficient quantity of
Electricity from the wire, and confequently
it was not immediately repelled. When the
air within the receiver was exhaufted above
one hundred times, the thread was not firft
attracted and then repelled as before, but
only vibrated a little backwards and forwards, and then remained in the fituation in
which it ftood when Electricity was not concerned.   By exhaufting the receiver ftill
farther, the vibration of the thread when
electrified was gradually diminifhed; fo that
when the degree of rarefaction was above
five hundred, fparks and the difcharge of a
jar only made the thread vibrate in a manner juft fenfible; but this vibration, however
fmall, did never become quite infenfible,
even when the receiver was exhaufted to the
utmoft power of the pump, which was very
near one thoufand. After this the air was gradually admitted into the receiver, and at various intervals the ball of the brafs wire was
electrified, in order to obferve whether the
fame phenomena appeared at the different
degrees

degrees of exhauſtion as had done before;
and they were found to agree with ſufficient
exactneſs.

EXPERIMENT II.

The braſs wire within the ſame glaſs receiver was made very ſhort, and from its extremity, a fine linen thread, ſix inches long,
was ſuſpended; and upon the plate of the
pump a ſmall braſs ſtand with a braſs pillar
was placed; ſo that when the receiver was
put upon the plate, and over the braſs ſtand,
about one inch length of the thread ſtood
parallel to, and at various required diſtances
from, the braſs pillar *. In this diſpoſition
of the apparatus, whenever any the leaſt
quantity. of Electricity was communicated
to the knob of the braſs wire, the thread was
immediately attracted by the braſs pillar, and
adhered to it ſome time, becauſe, being dry,
it did not immediately part with the acquired
Electricity. At various degrees of exhauſtion,
the electricity being communicated to the
braſs ball of the wire, it was found, that the

* This diſtance was altered by turning the braſs wire
which paſſed through a collar of leather in the braſs
cap of the receiver.

thread

thread was always attracted by the brafs pillar, though from a greater or lefs diftance, according as a greater or lefs quantity of air remained within the receiver. Thus when the air was rarefied about one hundred times, the thread was attracted from about one inch; when the air was rarefied two hundred times, it was attracted from about $\frac{1}{4}$ of an inch; when the air was rarefied three hundred times, it was attracted from about $\frac{1}{10}$; and after this it was always attracted from about one twentieth, even when the air within the receiver was rarefied about one thoufand times. It is remarkable, that when the air in the receiver is rarefied about three hundred times, if a jar is difcharged through the vacuum, by touching its knob with the ball of the wire on the receiver, the thread is not in confequence of it attracted by the brafs pillar; the reafon of which feems to be, becaufe that large quantity of Electricity opens at once a way through the vacuum, and paffes through every part of it; whereas a fmall quantity of Electricity, even the action of a fmall electrical machine in the fame room, at no very great diftance from the

apparatus,

apparatus, will caufe the thread being attracted by the brafs pillar.

EXPERIMENT III.

The brafs ftand, with the pillar, and the
thread which proceeded from the wire, being
removed from under the receiver, a very fenfible electrometer was faftened, inftead of the
thread, to the extremity of the brafs wire.
This electrometer confifted of two very fine
filver wires, each about one inch long, and
having a fmall cone of cork at its extremity. The fenfibility of fuch an electrometer
is really furprifing; for even the Electricity
of a fingle hair excited, does fenfibly affect
it; and, as its fufpenfion is almoft without
any friction or other impediment, it never
deceives one by appearing to be electrified
when in reality it is not fo. With this
preparation, the receiver being placed upon
the plate of the air-pump, the air was gradually exhaufted, and at intervals fome
Electricity was communicated to the ball
on the outfide of the receiver, either by an
excited electric or by a charged jar, and it
was found that the corks of the electrome-

ter were always made to diverge by it, even when the air was exhausted as much as possible. Indeed their divergency was smaller and smaller, and lasted a shorter time, according as the air was more exhausted, but it was visible to the last.

In this experiment, analogous to what has been observed in the preceding, when the air was exhausted above three hundred times, if a jar was discharged through the vacuum, or a strong spark was given to the knob on the top of the receiver, the corks of the electrometer diverged very little indeed, and but for an instant; whereas a small quantity of Electricity made them diverge more, and remain much longer in that state.

It seems deducible from those experiments, that electric attraction and repulsion take place in every degree of rarefaction, from the lowest to about one thousand, but that the power diminishes, in proportion as the air is more and more rarefied; and by following the low, we may perhaps conclude with F. Beccaria, that there is no electric attraction nor repulsion in a perfect va-

5   cuum:

cuum : though this will perhaps be impoffible to be verified experimentally, becaufe
when in an exhaufted receiver no attraction
or repulfion is obferved between bodies to
which Electricity is communicated, it will
be only fufpected, that thofe bodies are not
fufficiently fmall and light. But if we confult reafon, and which alone ought to affift
us when decifive experiments are not prac-
ticable, it feems likely that electric attraction and repulfion cannot take place in a
perfect vacuum, by which I only mean a
perfect abfence of air; becaufe either this
vacuum is a Conductor or a Non-conductor
of Electricity. If a Conductor, and as much
nearer to perfection as it becomes more free
from air, it muft be a perfect Conductor at
the fame time that it becomes a perfect vacuum, in which cafe electric attraction or
repulfion cannot take place amongft bodies
inclofed in it ; for, according to every notion
we have of Electricity, thofe motions indicate or are the confequence of the interven-
ing fpace in fome meafure obftructing the
free paffage of the electric fluid. And if
the perfect vacuum is a perfect Non-conductor, then neither electric attraction nor repulfion can happen in it.

H 2

EXPE-

EXPERIMENT IV.

In my former experiments having always obferved the electric light in the receiver of the air-pump, even when the air was rarefied to the utmoft power of that machine, I thought proper to repeat that experiment with receivers of various fizes; and accordingly have ufed receivers of above two feet in height, and fome of as large a diameter as the plate of the pump could admit, which is about fourteen inches, but the light in it was always vifible, only with different colours in different degrees of exhauftion, and always more diffufed and at the fame time lefs denfe when the air was more rarefied, which feems to render probable, that, when the air is quite removed from any fpace, the electric light is no longer vifible in it, as it muft have been the cafe with the experiment of Mr. WALSH's double barometer; for it is a maxim very well eftablifhed in Electricity, that the electric light is only vifible when the electric fluid, in paffing from one body to another, meets with fome oppofition in its way; and according to this propofition, when the air is entirely removed from a
given

given receiver, the electric fluid paffing
through that receiver cannot fhew any light,
becaufe it meets with no oppofition ; but
this will not account for the receiver ever
becoming a Non-conductor.

Having juft mentioned, that according as
the air is more and more rarefied in a receiver,
fo the electric light becomes gradually more
faint, it will be proper to add, that the electric light is more diffufed and lefs bright in
an exhaufted receiver than in air: thus, when
the receiver is not exhaufted, the difcharge
of a jar through fome part of it will appear
like a fmall globule exceedingly bright, but
when the receiver is exhaufted, the difcharge
of the fame jar will fill the whole receiver
with a very faint light; whereas fome perfons, by feeing the whole receiver illuminated,
are apt to fay that the light of Electricity is
rendered ftronger and greater by the exhauftion.

EXPERIMENT V.

It is mentioned by Mr. NAIRNE, in the
67th vol. of the Philof. Tranf. that having

H 3   put

put a piece of leather, juſt as it comes from the leather-ſellers, into the receiver of an air-pump, and afterwards having rarefied the air in it one hundred and forty-eight times, the electric light appeared very faint in it; whereas, without the leather, and even when the air was much more rarefied, the light of the electric fluid, when made to paſs through the receiver, was much more apparent. In conſequence of this obſervation, I ſuſpected that a little moiſture in the receiver, or ſome other effluvia of ſubſtances, might perhaps prevent the appearance of the electric light in rarefied air, and with this view I began to put various ſubſtances ſucceſſively into the receiver, and after rarefying the air by working the pump, ſome electric fluid was made to paſs through the receiver.

When a piece of moiſt leather was put into the receiver the air could not be rarefied above one hundred times, and the electric light appeared divided into a great many branches; though at the ſame time another ſort of faint light filled up the whole cavity of the receiver.

When a linen rag, moiſtened with a mixture

ture of fpirit of wine and water, was put into the receiver, the pump could not exhauft above forty times, and the light of Electricity appeared divided into many branches.

A wine-glafs full of olive oil placed under the receiver, prevented very little the exhauftion of the pump, the air being rarefied above four hundred times. The electric light appeared exactly as it ufually does in the fame degree of rarefaction when no oil is under the receiver, *viz.* a uniform faint light inclining to purple or red.

Concentrated vitriolic acid placed in a glafs under the receiver, produced no particular effect. 'As for the other mineral acids, they were not tried, becaufe, being volatile, they would have damaged the pump.

Dry folids, that had a confiderable fmell, as fulphur, aromatic woods previoufly made very dry, and fome refins, produced no particular effect, any more than fome of them prevented a very great degree of exhauftion, owing to fome moifture which ftill adhered to them.

H 4   From

From thefe experiments it appears, firft, that in the utmoft rarefaction that can be effected by the beft air-pump, which amounts to about one thoufand, both the electric light and the electric attraction, though very weak, are ftill obfervable; but, fecondly, that the attraction and repulfion of Electricity become weaker in proportion as the air is more rarefied, and in the fame manner the intenfity of the light is gradually diminifhed.   Now by reafoning on this analogy we may conclude, that both the attraction and the light will ceafe in a perfect abfence of air; but this will never account for this perfect vacuum ever becoming a Non-conductor of Electricity; for fince the electric fluid is very elaftic, and expands itfelf with more and more freedom in proportion as the refiftance of the air is removed, it feems unnatural that it fhould be incapable of pervading a perfect vacuum : however, the fact feems to be fully afcertained by Mr. Walsh and Mr. Morgan, and the only thing that remains to be done is to inveftigate the caufe of fo remarkable a property.

PART

PART V.

THE PRACTICE OF MEDICAL ELECTRICITY.

CHAP. I.

General Remarks relating to Medical Electricity.

THE wonderful effects of that unknown caufe generally named Electricity, foon after the difcovery of the electrical machines, were applied as a remedy for various diforders incident to the human body. The firft hints of this application, feem to have been fuggefted by obferving the effects produced upon thofe perfons that were electrified for curiofity; who being generally afraid of that extraordinary power, attributed entirely to it all thofe effects, which might in great meafure have been attributed to fear and apprehenfion: fuch were an increafed perfpiration, heat, increafe of pulfation, &c. The number of patients that were electrified at that time is prodigious, and the pretended cures

effected

effected by it were wonderful indeed. Accounts of thofe miracles performed by Electricity, were publifhed in various parts of Europe, together with the methods of electrifiying the patients; to which were added, fuch theories as, allowance being made for the infancy of Electricity at that time, would feem impoffible ever to have been propofed to the public. Thofe theories were ufually enforced by the account of experiments, which often proved falfe upon examination*. Indeed, if electrical machines could not be procured at prefent, we could hardly entertain any doubt concerning the veracity of thofe accounts, which had all the appearance of authenticity. But at prefent a much better acquaintance with the fcience of Electricity, than philofophers had about thirty or forty years ago, and lefs faith in the accounts of the generality of thofe perfons, whofe intereft it is to promote the ufe of Electricity in medicine; has pointed out the effects of that power upon the human body, in various circumftances, and has

---

* The medicated cylinders for electrical machines, are a remarkable inftance of this kind. See Dr. PRIESTLEY's Hiftory of Electricity.

fhewn

shewn how far we may confide in it; establishing, upon indisputable facts, that the power of Electricity is neither that admirable panacea it was 'confidered by fome fanatical and interefted perfons, nor fo ufelefs on application as others have afferted ; but that, when properly managed, it is an harmlefs remedy, which fometimes inftantaneoufly removes divers complaints, generally relieves, and often perfectly cures various diforders, fome of which could not be removed by the utmoft endeavours of phyficians and furgeons.

When the firft rumour occafioned in Europe by the accounts of pretended, and of a few real wonders, performed by means of Electricity, had in fome meafure fubfided, many creditable and experienced phyficians, who, juftly confidering it as their duty, had undertaken to examine the power of this new remedy, publifhed fome unfuccefsful applications of Electricity in divers difeafes ; in which cafes, they had not only prefcribed the electrization, but the operation had been performed either by themfelves, or under their infpection. · Thefe publications
gave

gave a new turn to the reputation of medical Electricity; and fince that time, the generality of phyficians and furgeons had not the leaft regard for its medical application; fo that the practitioners of it were rather confidered as fanatics and impoftors. However, an attentive examination of this fubject, after feveral trials, and after overcoming in great meafure the rooted prejudice amongft phyficians, began to eftablifh anew the reputation of medical Electricity; and fhewed that many applications of Electricity, publifhed in the above-mentioned accounts, had proved unfuccefsful, becaufe the operation was not managed properly; fo that it had been the abufe, and not the ufe of Electricity, that had proved unfuccefsful, and in fome cafes even detrimental; for at that time, ftrong fhocks and ftrong fparks were generally adminiftered, which a long feries of experiments and obfervations has proved to be generally ufelefs or hurtful. Mr. Lovet, who practifed medical Electricity for a long time, was, as far as I know, the firft who protefted againft the ufe of ftrong fhocks; and in an effay of his, intituled, *Subtil Medium proved*,

afferts,

afferts, that the fhocks to be ufed in medical Electricity fhould be very fmall; by which treatment he hardly ever failed of curing, or at leaft relieving his patients.

Electricity, different from other phyfical applications, requires rather a nicety of operation than a thorough knowledge of the difeafe. That it is poffible to apply Electricity properly, without a juft knowledge of the diforder, may feem a paradox; but it will be prefently fhewn, that to electrify a found part of the body together with the difeafed one, is by no means prejudicial, and that the degree of electrization muft be regulated rather by the feeling of the patient, than by the fpecies of diforder; from whence it muft follow, that the application of medical Electricity may be properly managed even with a fuperficial knowledge of the diforder. It muft, however, be confeffed, that farther experience may poffibly fhew much eafier and more certain methods of applying it differently for different difeafes; and therefore it is more likely that medical Electricity will receive improvements in the hands of fkilful phyficians

phyficians or furgeons, than when managed by ignorant perfons, whofe fuccefs is en-tirely trufted to chance.

The fuperiority of Electricity over other remedies, in many cafes, may appear from confidering, that medicines in general can-not always be confined to a particular part of the body, and to let them pafs through other parts is often dangerous, for which reafon they cannot ·be ufed; befides, after that thofe medicines have exerted their required power, they are with great diffi-culty, if at all, feparated from the body. But it is of no confequence whether the power of Electricity paffes through this or that other part of the body in order to come at the feat of the difeafe; and after having exerted its action, it is inftantly difperfed: hence it appears why Electricity has often . cured fuch obftinate diforders as have not yielded to any other treatment.

Having in the ninth Chapter of Part I. given a fummary view of the theory of Electricity, I fhall here only mention a few hints, which may promote the invefti-

gation

gation of the action of the electric fluid,
especially relating to its chymical action;
viz. if it adds any principle to those parts
through which it passes, as an acid, an al-
kali, the inflammable principle, &c. — The
observations relating to this point are, first,
that when any part of the body has been
exposed to the stream of electric fluid, it acquires a sulphurous, or rather a phosphoric
smell, which it retains for a considerable
time. Secondly, when the stream of electric fluid, issuing from a point, is directed
towards the palate, a kind of acid taste is
perceived. Now this smell and taste indicate, that the electric fluid either alters the
parts of the body, upon which it excites
those sensations, or that it carries along with
itself some other principle, which may perhaps be separated from those substances,
through which this fluid passes, previous to
its impinging upon the body.—Whether
those effects may be increased, diminished,
or turned to any use, and also whether they
are quite indifferent with respect to medical
Electricity, are matters that require farther
experiments and considerations; for nothing
certain has been yet determined respecting
- them.

In

In various experiments, when the electric
fpark is taken in air, or other fluids, efpecially the tincture of certain flowers, it fhews
effects fimilar to thofe, which the inflammable principle, or an acid, produces upon
thofe fluids *. Thefe experiments have induced various perfons to fuppofe, that the
electric fluid is phlogifton, or an acid, or
elfe a compound of both. But, confidering
that in thofe cafes the action of the electric
fluid as an acid, or as phlogifton, is exceedingly fmall; and alfo confidering the vio-
lence with which it paffes through the fubftance of bodies, the furface of which it
generally burns or melts in a fmall degree;
it feems more natural to fufpect, that the
above-mentioned effects are produced by
that quantity of inflammable or acid principle, which the violent paffage and efcape
of the electric fluid detaches from other
bodies, rather than to confider the electric
fluid itfelf to be an acid, or the inflammable
principle; which feems to be very unlikely
on various other accounts befides.

* See the laft four experiments of the Third Part.

A book

A book entitled *De l'eleЁricité du Corps humain, dans l'etat de fanté et de maladie, par l'Abbé Bertholon*, was publifhed a few years ago in France. In this work, the author, confidering the ufual Electricity of the atmofphere, imagines that the human body continually abforbs that Electricity of the air, and that this abforption is made through the pores of the fkin, as well as through the lungs in the ufual act of refpiration; and, as the quantity of air which enters into the lungs of a man in the courfe of one day, is by the author eftimated to about 1,152,000 cubic inches, he thence deduces, that the quantity of Electricity thus abforbed by the human body is aftonifhingly great. The author fucceffively examines the influence of this abforbed Electricity upon the functions of the body, *viz.* upon mufcular motion, upon the circulation of the blood, upon refpiration, digeftion, fecretions, and even upon the morals of men. He alfo takes notice of the Electricity of feveral animals, and even mentions fome qualities of the air, or of the aliments, which are proper to augment or diminifh the Electricity of the human body. In the

VOL. II.   I   application

application of thofe principles to medicine, the author gives a table of difeafes, arranged in fpecies and genera, and comprehended under ten claffes; which difeafes, as he fays, are occafioned in great meafure, if not wholly, by Electricity, either pofitive or negative; and which may be cured by one or the other of thofe two electrical powers, according as the figns concomitant the difeafe may indicate.

The Abbé Bertholon alfo treats of the influence of atmofpherical Electricity upon the number of births cr deaths, and other things of the like nature.

A perfon verfed in Electricity, confidering that the air next to the body of a man, in his ufual mode of living, is feldom if ever fenfibly electrified, and alfo that the human body is a ready conductor of Electricity, befides many other obvious confiderations, muft naturally fufpect that the author of the above-mentioned work has indulged his fancy perhaps too much.

Directions

CHAP. II.

Directions for the practical Application of Electricity for the Cure of various Diseases.

OMITTING the description of the electrical machine, and the manner of preserving it in good order (which things are to be found in the first part of his work) I shall only observe, with respect to the electrical machine in general, that its size should not be so small as was thought sufficient some time ago, when the smallest machines were supposed to be sufficiently useful for the purpose. It is somewhat remarkable, that when a small power of Electricity is to be used, large machines should be recommended; whereas, a short time ago, strong shocks were administered, and small electrical machines were used; but it must be considered, that when shocks are given, very small electrical machines can charge a Leyden phial much stronger than required; but when the stream is used, which has lately been found to be far more efficacious, then the small machines are

I 2 mostly

moftly ufelefs. Probably the largeft machines will not be found to afford a ftream too ftrong for medical purpofes ; but the ufeful ones, which do not require a great labour to be put in motion, and may furnifh a ftream fufficiently denfe, fhould have the glafs globe or cylinder about nine inches in diameter, which, with a proportionate Conductor, may ufually give fparks about three inches long. Whether the rubber of thefe machines ftands upon a glafs pillar or not, *viz.* whether it may be occafionally infulated or not, feems to be immaterial with refpect to medical Electricity : but as to have it fituate upon a glafs pillar is ufeful for electrical experiments in general, and perhaps it may be found hereafter, that negative electrization is beneficial in fome diforders, a perfon who is to choofe a new electrical machine, may rather have the rubber fixed upon a glafs pillar, than otherwife. Mr. NAIRN's new machine has every neceffary advantage *.

With fuch machines, the power of Electricity fhould be fo regulated, as to appl

, * See vol. I. p. 165.

ever

every degree of it with facility and readinefs; beginning with a ftream iffuing out
of a metal point, next ufing a wooden
point, then fmall fparks, ftronger fparks,
and laftly fmall fhocks.   Every one of
thefe methods may be increafed or diminifhed confiderably, by a proper management; thus, by turning the wheel of the
machine fwifter or flower, the ftream of
electric fluid may be regulated according
as the circumftances may require.   The
fparks may alfo be made ftronger or weaker, by taking them at a greater or lefs diftance, and by turning the wheel fwifter or
flower; and fo of the reft.

It is impoffible to prefcribe the exact
degree of electrization that muft be ufed
for various diforders; for perfons of different conftitutions, although afflicted with
the very fame difeafe, require different degrees of electrization.   Some perfons are of
fo delicate and irritable a conftitution, that
the fmalleft fparks give them as much pain
as fhocks do to others.   On the contrary,
fome people can fuffer pretty fevere fhocks
without pofitive pain; and I have heard,

though

though never faw any, of perfons who were infenfible of any electric power, even of confiderably ftrong fhocks.

In refpect to this important point, the operator muft be certainly inftructed by experience ; however, in the beginning, he may be affifted by the two following rules. Firft, he fhould begin to adminifter to his patients the very fmalleft degree of electric power, which he ought to continue for a few days, fo as to obferve whether it produces any good effect, which if it fails to do, he fhould then increafe the ftrength of Electricity ; and fo proceed gradually till he finds the effectual method, which he fhould follow without variation, till the patient is intirely cured. In fhort, the operator fhould always ufe the fmalleft degree of electric power, that is fufficient for the purpofe. A little practice will enable him to determine at once what degree of Electricity is required for his patient, without any ufelefs attempts. Secondly, the degree of electrization to be adminiftered, fhould never exceed that, which the patient can conveniently fuffer;

experience

experience fhewing, that when the application of any degree of Electricity is very
difagreeable to the patients, they very feldom
mend.

The inftruments which, befides the
electrical machine and its prime Conductor, are neceffary for the adminiftration of
medical Electricity, may be reduced to
three ; *viz.* an electric jar, with Mr. LANE's
electrometer ; an infulated chair, or an infulated ftool, upon which a common chair
may be occafionally fet ; and the directors *.

Thofe inftruments are delineated in
the fifth plate. Fig. 1. reprefents the
electric jar, with Mr. LANE's electrometer, and the manner in which the
fhocks are fent through any particular part
of the body. The furface of the jar, which
is coated with tin-foil, fhould be about four
inches in diameter, and fix inches high,
which is equal to about feventy-three

* Various other inftruments ufeful in medical electricity, are defcribed in divers books, but thofe mentioned above are fufficient to anfwer every required
purpofe.

fquare

square inches. The brafs wire, which paffes through the covering of the jar and touches the infide coating, has a brafs ball, F, to which the electrometer F D E is faftened; and proceeding a little farther up, terminates in another brafs ball B, which fhould be fo high as to touch the prime Conductor A, which is fuppofed to ftand before the electrical machine. The electrometer confifts in a glafs ftick F D, cemented to two brafs caps F and D; from the latter of which a ftrong perpendicular brafs wire proceeds, the extremity of which comes as high as the center of the ball B, and is furnifhed with an horizontal fpring focket, through which the wire C E, having the brafs ball C at one end, and the open ring E at the other, may be flided backwards and forwards, fo as to fet the brafs ball C at any required diftance from the ball B. This diftance, at moft, needs not be greater than half an inch; hence the electrometer may be made very fmall. Sometimes fmall divifions are marked upon the wire C E, which ferve to fet the balls B and C at a given diftance from one another, with more readinefs and precifion. Now fup-

poſe

pofe that the jar is fet contiguous to the prime Conductor, that is, with the ball B touching the Conductor; that the ball C be fet at one tenth of an inch diftance from the ball B; and that, by means of wire, a conducting communication be formed from E to the outfide coating of the jar, as is reprefented by the dotted line in the figure. In this cafe, if the electrical machine be put in motion, the jar will be charged; and when the charge is fo high as that the electric fluid accumulated within the jar can leap from the ball B to C, which we have fuppofed to be one tenth of an inch afunder, the difcharge will happen, a fpark appearing between the faid balls, and the fhock paffes through the wire reprefented by the dotted line; for the part F D of the electrometer being of glafs, generally covered with fealing-wax, is impervious to Electricity, confequently the electric fluid has no other way through which it can pafs from the infide to the outfide of the glafs jar. When the fhocks are to be given with this apparatus to any particular part of the body, for inftance, to the arm, then, inftead of the dotted line reprefenting a wire, which muft

now

now be fuppofed as not exifting in the figure, two flender and pliable wires, E L, I L, are to be faftened, one to the open ring E of the electrometer, and the other to the brafs hook I of the ftand H I, which communicates with the outfide coating of the jar *. The other extremities of the faid wires are faften- ed each to the brafs wire L, and L, of the directors K L, K L. Each of thofe inftru- ments, juftly called *directors*, confifts of a knobbed brafs wire L, which by means of a brafs cap is cemented to the glafs handle K. The operator, holding them by the extremity of the glafs handle, brings their balls into contact with the extremities of that part of the body of the patient through which he defires to fend the fhock. The management and convenience of this appa- ratus are eafily comprehended by infpecting the figure; for when the machine is in mo- tion, and the apparatus, &c. is fituate as in the figure, the difcharge of the jar muft be

---

* If the jar has not the ftand H I, the extremity I of the wire I L may be fimply refted under, or may be tied round it. In fhort, it muft be put in contact with the outfide coating of the jar, in any convenient manner.

evidently

evidently made through that part of th
patient's arm, which lies between the
knobs of the directors ; and the operator,
whilft an affiftant keeps the machine in
motion, has nothing more to do, than to
hold the knobs of the directors to the
extremities of the arm, or to any other
part of the body that is required to be thus
electrified ; always taking care that the two
wires E L, I L, do not touch each other,
becaufe in that cafe the fhock will not pafs
through that part of the body which is required to be electrified.   Thus any number
of fhocks, precifely of the fame ftrength,
may be given, without altering any part of
the apparatus, or having any farther trouble;
and when the ftrength of the fhocks is required to be diminifhed or increafed, it is
only neceffary to diminifh or augment the
diftance between the balls B C, which is
done by flipping the wire C E forwards or
backwards through the fpring focket that
holds it.

It is almoft fuperfluous to mention, that
when fhocks are adminiftered, it is immaterial whether the patient ftands upon the
ground,

ground, upon the infulating ftool, or in any
other fituation whatever. It is neither
always neceffary to remove the cloaths from
the part that muft be electrified, in order
to let the knobs of the directors touch the
fkin; for, except the coverings be too many
and too thick, in which cafe part of them
at leaft fhould be removed, the fhocks will
go through them very eafily, efpecially if
the knobs of the directors be preffed a little
upon the part.

In the courfe of this effay we fhall defcribe the ftrength of the fhocks by the
diftance between the balls B and C of the
electrometer, which we fhall exprefs by
parts of an inch; fuppofing that the faid
electrometer is fixed upon fuch a jar as we
have defcribed above, *viz.* whofe coated
part, befides the bottom, may be equal to
about 73 fquare inches, and whofe glafs is
moderately thin; for a larger or thicker jar
with the fame electrometer, fet at the fame
diftances, will produce a much different
effect, as muft be obvious to any perfon a
little acquainted with the fcience of Electricity.

Befides

Befides the directors mentioned above, there are other kinds of directors, which ferve for throwing the ftream of electric fluid, and other fimilar purpofes. Thefe are delineated in fig. 2. and 3. The director D, in fig. 2. is much like thofe defcribed above, excepting only that its wire is bent, and inftead of having any ball, it terminates in a point, to which is affixed a piece of wood about one inch or one inch and a half long, pointed on one end, though not very fharp, and having a hole on the other *. The operator fhould have by him various fuch wooden pieces, of different length and thicknefs, as E E, fo as to fhift them according as circumftances may require ; for fometimes the wooden pieces are too dry or too damp, or the machine is in bad order, &c. in which cafes the ftream of electric fluid would be either too ftrong or too weak, if the fame wooden point was always ufed. The wood proper to make thefe pointed pieces fhould be rather of a foft kind, than hard, as box wood and lignum vitæ are.

* Thefe directors are fometimes made with very flender and annealed wires, fo that they may be bended in every required direction.

In

In order to throw the electric fluid with this director, let a wire B, proceeding from the prime Conductor, A, fig. 2. be faftened to the wire of the director D E, which the operator muft 'hold by the extremity of the glafs handle, and muft manage it fo as to keep the wooden point at about one or two inches diftance from the body of the patient *. This diftance, however, muft be regulated according to the conftitution of the patient, the ftrength of the electrical machine, and other circumftances, which will be fuggefted by a little practice. The electric fluid iffuing from the wooden point, has a power which is intermediate between that of the ftream proceeding from a metal point, and the power of the fparks; but yet it is in general the moft efficacious method of electrization, and therefore no pains fhould be fpared in order to adminifter it in the beft poffible manner. This ftream confifts of a vaft number of exceedingly fmall fparks, accompanied with a little

* When this or any other operation is performed, the electric jar, and in general any inftrument not actually neceffary, muft be removed from the prime Conductor, and even from the table if that is rather fmall.

wind,

wind, which gently irritates the part electrified, and gives a warmth which proves
very agreeable to the patients. Sometimes,
when the machine is very powerful, and the
wooden point is fhort or fplit, a very full
and pungent fpark iffues from it; which is
a very difagreeable accident, efpecially when
the part electrified is very delicate. In order
to avoid this inconvenience, the operator
fhould firft try the goodnefs of the point,
before he begins the operation; which he
may do by throwing the ftream upon his
own hand or face.

The above-mentioned method of electrifying, gentle as it may appear, will ne-
verthelefs be found too ftrong for fome perfons, efpecially when ufed for open fores
upon delicate parts; in which cafes the
wooden point muft be removed, and the
electric fluid muft be fimply thrown
from the metal point of the director, which
muft now be kept at a greater diftance
than when the wooden piece was upon it.
The electric fluid iffuing out of this pointed
wire of the director, occafions only a gentle wind upon the part towards which it

6   is

is directed, and is far from being difagee-
able even to the moft delicate conftitution.

It might be naturally fufpected, that fo
gentle and nearly infenfible a treatment
could hardly be of any efficacy; but my
reader may be affured, that to my certain
knowledge, deduced from the practice of
perfons who have had long experience in
this fubject, this method of electrization,
viz. the throwing the fluid with a metal
point, has often mitigated pains, and cured
obftinate and dangerous difeafes, which
could not be removed by any other remedy
that was tried.

In general this treatment, upon delicate
nervous conftitutions, is as efficacious as
the other, *viz.* the throwing the fluid with
a wooden point, is to ordinary conftitutions.
In feveral cafes, efpecially of open fores,
the electric fluid iffuing out of a wooden
point has conftantly increafed the pain, and
even enlarged the fore; whereas the fluid
iffuing out of the metal point has effectually diminifhed both.

The

The ftream iffuing out of a wooden point may be directed towards the eyes of the patient, without any apprehenfion of hurting him; in which cafe the operator fhould keep the eye-lid open with one hand. Indeed there might be fome cafes, though I feldom heard of any, in which this treatment may be thought to be too ftrong; then the metal point only may be ufed.

The ftream iffuing both out of th wooden and of the metal point, acts even through the cloaths, if they are not too thick; hence it may be ufed without in-commoding the patient; but when it is convenient to uncover the part that is to be electrified, it is much preferable to direct the fluid immediately upon the fkin.

In this operation, the practitioner muft mind to fhift the point of the director about, fo that the ftream of electric fluid may be directed not only towards the affected part, but alfo to the places about it; alternately returning to the fame place, and moftly infifting upon the part princi-pally affected.

VOL. II.   K   The

The patient in this operation may also ftand in every fituation that may happen to be more convenient to him.

When more proper inftruments cannot be had, directors may be made by fticking large pins upon fticks of fealing-wax, as is reprefented at K, fig. 2.

Sometimes the wire B, which forms the communication between the prime Conductor and the director, throws a confiderable quantity of electric fluid into the air, which weakens the ftream iffuing from the point. In order to remedy this inconvenience, I contrived a conducting wire, which being ufed by fome of my friends, who practife Medical Electricity, has been found to anfwer very well the purpofe of not diffipating the electric fluid. This conducting communication is formed of a filver, gold, or copper thread, fuch as are ufed for laces, which confift of a fmall lamina of metal twifted round a filk or linen thread. This metal thread, or two of them, I involve in a filk ribbon, which is coiled and fewed very tight upon it, leaving only a loop of the metallic

ietallic thread uncovered at each extremity,
ie of which is to be faftened to the prime
onductor, and the other to the wire of the
irector.   See G H, fig. 2.

This fort of conducting communication,
efides its preventing the diffipation of the
ectric fluid, is much more pliable than
ie ftiff wire commonly ufed, and confe-
uently may be managed more eafily.
: may be alfo ufed inftead of the wires
L, I L, fig. 1. in the operation of giving
iocks.

Two other directors different from the
bove-mentioned, are delineated in fig. 3.
Their ufe is to draw fparks from the in-
de of the ear in cafes of deafnefs, pains,
cc. and alfo from the teeth or other inter-
al parts of the mouth.   The director B H
onfifts of a glafs tube A B, about fix
nches long, and open at both ends; the dia-
neter of which may be about one tenth of an
nch, and the fubftance of the glafs rather
hick.   A cork is thrufted into one end of
his tube, through which a wire paffes; one
xtremity of which is cut blunt and fmooth,

K 2   and

and comes within one or two tenths of an inch fhorter than the end B of the tube The other extremity H of the wire, i furnifhed with a fmall metal ball.—Long pins, fuch as the ladies ufe for their hats anfwer this purpofe exceedingly well, when their points are filed off. The other direc- tor C D differs from that juft defcribed, in being only bent a little, for the convenien- cy of adapting it more eafily to fome parts within the mouth.

When thefe directors are ufed, the pa- tient muft be fituated upon an infulating ftool, *viz.* a ftool furnifhed with glafs feet, upon which a chair may be placed. Then a communication muft be formed between the prime Conductor and the body of the patient, by means of any fort of wire, ef- pecially that reprefented by G H, fig. 2. or by the patient only touching the prime Conductor with his hand. In this cafe, it is eafy to conceive that the patient be- comes part of the prime Conductor; and if any blunt conducting body is brought near him, when the machine is in action, a fpark is obtained from him in the fame manner as when the fame blunt body is prefented to the prime Conductor itfelf. Every thing being thus far prepared, the operator holding the director A B or C D by its middle E or F with one of his hands, muft bring the extremity B or D of the tube into contact, or nearly. fo, with the infide of the ear, mouth, &c. of the patient, as occafion may require ; and muft bring the knuckle of a finger of his other hand within a fmall diftance of the fmall knob H or G of the director, which will extract fmall fparks from it, and at the fame time the like fparks will happen between the other extremity of the wire within the tube, and the part of the patient's body towards which the inftrument is directed.

This is an excellent method to be prac-tifed in cafes of deafneffes, pains in the ears, tooth-achs, fwellings within the mouth, &c. efpecially becaufe it may be increafed or diminifhed at pleafure ; *viz.* by drawing the wire G or H more or lefs from the extremity B or D of the tube, the ftrength of the fparks may be increafed or diminifhed.

K 3

Not

Not only fparks, but alfo the ftream of electric fluid may be drawn with thofe directors. This is done by bringing (inftead of the knuckle) a pointed piece of wood near the fmall knob H or G of the director; every thing elfe being difpofed as already directed.

When fparks are required to be drawn from any part of the body, the patient muft be fituated upon an infulating ftool, and muft be connected with the prime Conductor in the manner directed above; then the operator bringing the knuckle of one of his fingers, or the knob of a brafs wire like K L, fig. 3. oppofite to the affected part, will draw the fparks from it; which fparks will pafs very eafily through the cloaths, if they are not very thick. When the knobbed wire K L is ufed to draw fparks with, the operator muft hold it by the extremity K, and prefent the knob L, &c. : but it may alfo be ufed to draw the fluid filently, in which cafe the point K muft be prefented, and the knob L muft be held by the hand of the operator. Here a wooden point may alfo be

ufed,

ufed, *viz.* by affixing it to the point K of
the wire; which method anfwers as well
as that of throwing the electric fluid by
means of a wooden or metal point, with
the director D of fig. 2. defcribed in the
preceding pages.

Sometimes it is required to take fparks
from fuch parts as are covered with thick
cloaths, and the patient is rather unwilling
to uncover.   In this cafe the beft method
is to fituate the patient upon the infulating
ftool in contact with the prime Conductor,
then to bring the knob of a director, like
one of thofe delineated in fig. 1. in contact with the cloaths over the part required
to be electrified; whilft the operator, holding the inftrument by the extremity of its
glafs handle with one of his hands, brings
the knuckle of one of his fingers, or the
knob of the wire K L, fig. 3. pretty near
the brafs cap of the director, fo as to draw
ftrong fparks from it; the force of which
will be felt very fmartly upon the part of
the patient's body; for at the fame time
fparks will happen there acrofs the cloaths,
viz. between the part of the body of the
K 4   patient,

patient, and the knob of the director, which, for better fecurity, fhould be preffed a little upon the cloaths.

In all thofe cafes when the electric fluid, either in a ftream or under the form of fparks and fhocks, is to be forced acrofs the cloaths, it is fuppofed that no metallic ornaments fhould be interpofed, as gold or filver lace, long pins, and the like; for then the effects will vary confiderably, according to the different circumftances.

There is another method of electrifying a difeafed part of the body, which cannot properly be called drawing fparks, though it comes very near to it. This manner of electrifying is effected in the following way: The patient is fituated upon the infulating ftool, and is made to communicate with the prime Conductor; then a dry and warm flannel, either fingle or double, according as it may be occafionally thought more proper, is fpread upon the naked part that muft be electrified, and over this flannel the operator muft put the knob L of the wire K L, fig. 3. quite into contact

tact with the flannel, whilst he holds it by
the other extremity K. Now when the
machine is in action, the knob L of the
wire must be shifted very quick and nimbly
from place to place over the flannel; in
which manner a vast number of exceedingly
small sparks will be drawn across the flannel; which generally bring an agreeable
warmth on the part, and prove very beneficial to the patient, at the same time that
they do not cause any very disagreeable sensation. In cases of paralytic limbs, rheu-
matism, spreading pains, coldness of any
particular part, &c. this treatment is of sin-
gular benefit. In the following pages we
shall call it *The method of drawing sparks
through a piece of flannel,* or simply, *to draw
sparks through flannel.*

As for the insulating chair, it is almost
needless to give any particular directions
concerning its construction; it being nothing more than a common wooden chair
set upon an insulating stool; or, as some
persons choose to have it, the chair itself
is furnished with glass instead of wooden
legs, which answers equally well. It is
                                    requisite

requifite that no fharp metallic points be put upon this chair; and even its wooden ornaments fhould be rather blunt than fharp-edged; for points and edges in general diffipate the electric fluid confiderably, and confequently weaken the power of the machine. The glafs feet fhould be at leaft eight inches high; and, that they may infulate the better, efpecially in damp weather, they fhould be covered with fealing-wax or good amber varnifh. In the conftruction of this chair, a place fhould be always provided, whereupon the patient may reft his feet; the want of which is very difagreeable, becaufe it is abfolutely neceffary that the feet do not touch the floor.—When only a common chair is to be occafionally fet upon an infulating ftool, the latter fhould be made fomewhat larger than the former, fo that part of it may project before the chair, upon which the perfon to be electrified may reft his feet.

After the defcription of the inftruments neceffary for the adminiftration of Medical Electricity, I fhall collect together fome practical rules, which may ferve for a guide

to

to thofe practitioners, who have not yet been fufficiently inftructed by their own experience.

General Rules for Practice.

I. It fhould be attentively obferved, to employ the fmalleft force of Electricity, that is fufficient to remove or to alleviate any diforder; thus the fhocks fhould never be ufed when the cure may be effected by fparks; the fparks fhould be avoided when the required effect can be obtained by only drawing the fluid with a wooden point; and even this laft treatment ought to be omitted, when the fluid drawn by means of a metal point, may be thought fufficient. The difficulty confifts in dif-tinguifhing the proper ftrength of electric power that is required for a given difor-der, the fex and conftitution of the pa-tient being confidered. In regard to this point, it is impoffible to give any exact and invariable rules; the circumftances being of fuch a nature, and fo various, that long experience, and a ftrict attention to every particular phenomenon, are the only means by by which proper inftructions may be received. The fureft rule, as we obferved above, that can be given relating to this particular, is to begin by the moft gentle treatment; at leaft fuch, that, confidering the conftitution of the patient, may be thought rather weak than ftrong. When this gentle treatment has been found ineffectual for a few days, which is denoted by the difeafe not abating, and the application of Electricity not caufing any warmth, or other promifing phenomenon, upon the part electrized; then the operator may gradually increafe the force of Electricity till he finds the proper degree of it.

II. In judging of cafes proper to be electrified, experience fhows, that in general, all kinds of obftructions, whether of motion, of circulation, or of fecretion, are very often removed or alleviated by Electricity. The fame may alfo be faid of nervous diforders; both which include a great variety of difeafes. The application of Electricity has feldom intirely cured difeafes of a long ftanding, although it generally relieves them. To perfons afflicted with

the

the venereal difeafe, or to pregnant women,
electrization has been thought to be pernicious; but my reader may be affured, that
even in thofe cafes it may be ufed without
fear, if it is judicioufly managed. When
pregnant women are to be electrified for any
diforder, the fhocks fhould be abfolutely
forbidden; and even when the other more
gentle treatments are ufed, a conftant attention fhould be given to any phenomenon
that may appear in the courfe of the electrization; the method of which fhould be
increafed, diminifhed, or fufpended, according as circumftances may indicate. As for
the venereal difeafe, it will be hinted, in
the courfe of this work, in what manner,
and in which cafes, Electricity may be
applied.

III. In cafes of gathering tumours, the
beft method is to draw the fluid by means
of a wooden point, or, if that proves painful,
by a metal point. Sparks in thefe cafes,
and alfo fhocks, are often hurtful. In ftiff-
neffes, paralyfies, and rheumatifm, fmall
fparks, efpecially through a double flannel,
and alfo very fmall fhocks (at moft of one
tenth

tenth of an inch) may be ufed. Stronger
fhocks may be fometimes, though feldom,
adminiftered for a violent tooth-ach, and for
fome internal fpafm of no long ftanding.

IV.. When any limb of the body is deprived of motion, it muft be obferved, that
the privation of motion is not always origi-
nated by a contraction of the mufcles; but
that it is often occafioned by a relaxation;
thus, for inftance, if the hand is bent in-
wardly, and the patient has no power of
ftraightening it, the caufe of it may be a
weaknefs of the outward mufcles, as well
as a contraction of the inward ones. In
fuch cafes, as it is often difficult even for
good anatomifts to difcover the real caufe,
the fureft method is to electrify not only
thofe mufcles which are fuppofed to be
contracted, but alfo their antagonifts; for to
electrify a found mufcle is by no means
hurtful.

V. When the ftream of electric fluid is
thrown either with a wooden or metal
point, the length of the operation fhould
be from three to ten minutes: more or

lefs,

lefs, according as occafion may require. When fhocks are adminiftered, their greateft number fhould not exceed a dozen or fourteen, except when they are to be given to the whole body in different directions. The number of fparks, when they are ufed, may generally exceed the number of fhocks mentioned above.

VI. Laftly, it may be of ufe to mention, that when children muft be electrified upon the infulating chair, as it is difficult to let them ftay without motion, the moft convenient method is, to let another perfon fit in the infulating chair, and to hold the child whilft the operator is electrifying him.

Having thus comprifed into a few general rules, the method of applying Electricity with fafety, we fhall next defcribe the particular treatment, which has been found more expeditious and beneficial in diforders of various fpecies; and fhall laftly add fome authentic cafes, which will ferve as examples for the generality of practitioners.

The minute detail of the apparatus neceffary

ceffary for Medical Electricity, and feveral
of the obfervations made on the particular
ufe of it, may be thought more than neceffary
for a perfon who is mafter of the preceding
part of this work; but as it is more than
probable, that many perfons will confult
this book merely for the fake of the medical
part, I thought proper to prevent any poffible miftake, or infufficient inftruction,
which might arife from concifenefs.

C H A P. III.

Containing the particular Method of adminif-
tring Electricity for various Difeafes, and
the Account of fome authentic Cafes.

THE account of a few fuccefsful cafes
in Medical Electricity, as well as in
any other branch of phyfic, does by no
means eftablifh the reputation of the treatment, when a vaft number of unfuccefsful
trials are concealed from the eyes of the
public. The variety of temperaments obfervable in the human fpecies, and the coin-
cidence of circumftances, is fuch, that
fometimes very obftinate diforders feem to
be cured by very trifling applications. The

phyficians,

ɔhyſicians, however, juſtly neglect thoſe
inds of treatment, becauſe they have actu-
lly failed in a great many caſes ſeemingly
ɔf the ſame nature.

In order, therefore, to give a proper eſtimate of the efficacy of a remedy, it is neceſſary to ſhew the proportion between the
ſucceſsful, and the unſucceſsful trials;
without being amazed at one caſe, and
neglecting many others.

Agreeably to this obſervation, the reader
will find in the following pages, an eſtimate
of the effects of Electicity applied as a remedy for various diſorders. This eſtimate
has been deduced from the caſes which are
hitherto come to my notice, and is therefore likely to receive much alteration and
amendment by better information, and future obſervations.

Rheumatic diſorders, even of long ſtanding, are relieved, and generally quite cured,
by only drawing the electric fluid with a
wooden point from the part, or by drawing ſparks through flannel. The operation ſhould be continued for about four

or five minutes, repeating it once or twice every day.

Deafneſs, except when it is occaſioned by obliteration, or other improper configuration of the parts, is either intirely or partly cured by drawing the ſparks from the ear with the glaſs-tube-director, or by drawing the fluid with a wooden point. Sometimes it is not improper to ſend exceedingly ſmall ſhocks (for inſtance, of one-thirtieth of an inch) from one ear to the other.—It has been conſtantly obſerved, that whenever the ear is electrified, the diſcharge of the wax is conſiderably promoted.

The tooth-ach, occaſioned by cold, rheumatiſm, or inflammation, is generally relieved by drawing the electric fluid with a point, immediately from the part, and alſo externally from the face. But when the body of the tooth is affected, electrization is of no uſe; for it ſeldom or never relieves the diſorder, and ſometimes increaſes the pain to a prodigious degree.

Swellings in general, which do not contain any *matter*, are moſtly cured by
drawing

drawing the electric fluid with a wooden point *. The operation fhould be continued for three or four minutes every day.

Inflammations of every fort are generally relieved by a very gentle electrization.

In inflammations of the eyes, the throwing of the electric fluid by means of a wooden point, is conftantly attended with great benefit; the pain being quickly abated, and the inflammation being generally diffipated in a few days. In thefe cafes, the eye of the patient muft be kept open, and care fhould be taken not to bring the wooden point very near it. Sometimes it is fufficient to throw the fluid with a metal point; for in thefe cafes, too great an irritation fhould be always avoided. It is not neceffary to continue this operation for three or four minutes without intermiffion; but, after throwing the fluid for about half a minute, a fhort time may be allowed to the patient to reft, and to wipe

---

* It is very remarkable, that in fome cafes of white fwellings, quite cured by means of Electricity, even the bones and cartilages were in fome meafure disfigured.

　　　　　　　his

his tears, which generally flow very copioufly; then the operation may be continued again for another half minute, and fo on for four or five times every day.

The *gutta ferena* has been often cured by electrization; but at the fame time it muft be confeffed, that to my certain knowledge, Electricity has proved ineffectual in many fuch cafes, in which it was adminiftered for a long time, and with all poffible attention. I do not know that ever any body was worfted by it. The beft method of adminiftring Electricity in fuch cafes, is firft to draw the electric fluid with a wooden point for a fhort time, and then to fend about half a dozen fhocks of one-twentieth of an inch from the back and lower part of the head to the forehead, very little above the eye.

A remarkable difeafe of the eyes was fome time ago perfectly cured by electrization; it was an opacity of the vitreous humour. This feems to be the only cafe of the kind, to which Electricity was applied.

All

All the cafes of *fiftula lacrymalis,* as far
as I am informed, that have been electrified
by perfons of ability for a fufficient time,
have been entirely cured. The method
generally practifed, has been that of drawing the fluid with a wooden point, and to
take very fmall fparks from the part. The
operation may be continued for about three
or four minutes every day. It is remarkable
that in thofe cafes, after curing the fiftula lacrymalis, no other difeafe was occafioned by it, as blindnefs, inflammations,
&c. by fuppreffing that difcharge.

Palfies are feldom perfectly cured by
means of Electricity, efpecially when they
are of long ftanding and the intellect is
affected ; but they are generally relieved to
a certain degree. The method of electrifying in thofe cafes, is to draw the fluid
with the wooden point, and to draw fparks
through flannel, or through the ufual coverings of the part, if they are not too thick.
The operation may be continued for about
five minutes *per* day.

Ulcers, or open fores of every kind, even

of

of a long ftanding, are generally difpofed to heal by electrization. The general effects are a diminution of the inflammation, and at firft a promotion of the difcharge of properly formed matter; which difcharge gradually leffens, according as the limits of the fore contract, till it is quite cured. In thefe cafes the gentleft electrization muft be ufed, in order to avoid too great an irritation, which is generally hurtful. To draw or throw the fluid with a wooden or even with a metal point, for three or four minutes *per* day, is quite fufficient.

Cutaneous eruptions have been fuccefsfully treated with electrization; but in thefe cafes it muft be obferved, that if the wooden point is kept too near the fkin, fo as to caufe any confiderable irritation, the eruption will fometimes be caufed to fpread more; but if the point be kept at about fix inches diftance, or farther, if the electrical machine is very powerful, the eruptions will be gradually diminifhed, till they are quite cured. In this kind of difeafe, the immediate and general effect of the wooden point, is to occafion a warmth about the

electrified

electrified part, which is always a fign that the electrization is rightly adminiftered.

The application of Electricity has perfectly cured various cafes of *St. Vitus's Dance,* or of that difeafe which is commonly called fo; for it is the opinion of fome very learned phyficians, that the real difeafe called St. Vitus's Dance, which formerly was more frequent than it is at prefent, is different from that which now goes under that name. In this difeafe, fhocks of about one-tenth of an inch may be fent through the body in various directions, and alfo fparks may be taken. But if this treatment proves very difagreeable to the patient, then the fhocks muft be leffened, and even omitted; inftead of which, fome other more gentle applications muft be fubftituted.

Scrophulous tumors, when they are juft beginning, are generally cured by drawing the electric fluid with a wooden or metal point from the part. This is one of thofe kinds of difeafes in which the action of Electricity requires particularly the aid of

L 4. other

other medicines, in order to effect a cure more eafily ; for fcrophulous affections generally accompany a great laxity of the habit, and a general cachexy, which muft be obviated by proper remedies.

In *cancers*, the pains only are moftly alleviated by drawing the electric fluid with a wooden or metal point. I know of one cafe only, in which a moft confirmed cancer of very long ftanding, on the breaft of a lady, has been much reduced in fize. It is remarkable, that this patient was fo far relieved by drawing the fluid with a metal point from the part, that the excruciating pains fhe had fuffered for many years, did almoft intirely difappear ; and alfo, that when the electric fluid was drawn by means of a wooden point, the pains did rather increafe. This perfon, when I heard of her laft, was ftill under the application of Electricity; and the cancer feemed not unlikely to be perfectly cured, although contrary to the expectations even of the judicious phyfician who electrified her, and who knows too well the nature of that dangerous difeafe.

Abfceffes,

Abfceffes, when they are in their beginning, and in general whenever there is
any tendency to form matter, electrization
difperfes them.   Lately, in a cafe in which
matter was formed upon the hip, called
the *lumber abfcefs,* the difeafe was perfectly
cured by means of Electricity.   The *fciatica*
has alfo been often cured by it.   In all fuch
cafes, the electric fluid muft be fent through
the part by means of two directors applied
to oppofite parts, and in immediate contact
either with the fkin, or with the coverings,
when thefe are very thin.   It is very remarkable, that the mere paffage of the
electric fluid in this manner, is generally
felt by the patients afflicted with thofe diforders, nearly as much as a fmall fhock is
felt by a perfon in good health.   Sometimes
a few fhocks have been alfo given, but it
feems more proper to omit them ; becaufe
fometimes, inftead of difperfing, they rather
accelerate the formation of matter.

In cafes of *pulmonary inflammations,* when
they are in the beginning, electrization
has fometimes been beneficial ; but in confirmed difeafes of the lungs, I do not know
that

that it ever afforded any unqueſtionàble benefit ; however, it ſeems that in ſuch caſes the power of Electricity has been but ſeldom tried.

Nervous head-achs, even of a long ſtanding, are generally cured by electrization. For this diſeaſe, the electric fluid muſt be thrown with a wooden, and ſometimes even with a metal point, all round the head ſucceſſively. Sometimes exceedingly ſmall ſhocks have been adminiſtered ; but theſe can ſeldom be uſed, becauſe the nerves of perſons ſubject to this diſeaſe are ſo very irritable, that the ſhocks, the ſparks, and ſometimes even the throwing the electric fluid with a wooden point kept very near the head, throws them into convulſions.

The application of Electricity has often been found beneficial in the *dropſy,* when juſt beginning, or rather in the tendency to a dropſy ; but it has never been of any uſe in advanced dropſies. In ſuch caſes, the electric fluid is ſent through the part, in various directions, by means of two directors, and ſparks are alſo drawn acroſs

the

the flannel or the cloaths; keeping the metal rod in contact with them, and shifting it continually from place to place. This operation should be continued at least ten minutes, and should be repeated once or twice a day.—Perhaps in those cases, a simple electrization, (*viz.* to insulate the patient, and to connect it with the prime Conductor whilst the machine is in action) continued for a considerable time, as an hour or two, would be more beneficial.

The *gout*, extraordinary as it may appear, has certainly been cured by means of Electricity, in various instances. The pain has been generally mitigated, and sometimes the disease has been removed so well as not to return again. In those cases, the electric fluid has been thrown by means of a wooden point, although sometimes, when the pain was too great, a metal point only has been used.

Agues very seldom fail of being cured by Electricity, so that sometimes one electrization, or two, have been sufficient. The most effectual and sure method has been

that

that of drawing fparks through flannel, or the cloaths, for about ten minutes, or a quarter of an hour. The patients may be electrified either at the time of the fit, or a fhort while before the time in which it is expected.

The fuppreffion of the ordinary periodical fluxes of women, which is a difeafe that often occafions the moft difagreeable and alarming fymptoms, is fuccefsfully and fpeedily cured by means of Electricity, even when the difeafe is of long ftanding, and after the moft powerful medicines ufed for it have proved ineffectual. The cafes of this fort, in which electrization has proved ufelefs, are fo few, and the fuccefsful ones fo numerous, that the application of Electricity for this difeafe may be juftly confidered as an efficacious and pretty certain remedy. Great attention and knowledge is required, in order to diftinguifh the arreft of the periodical fluxes from a ftate of pregnancy. In the former, the application of Electricity, as we obferved above, is very beneficial; whereas in the latter, it may be attended with very difagreeable effects; it is therefore a matter

of

of great importance to afcertain the real caufe of the difeafe, before the Electricity be applied in thofe cafes. Pregnant women may be electrified for other difeafes, but always ufing very gentle means, and directing the electric fluid through other parts of the body, diftant from thofe fubfervient to generation. In the real fuppreffion of the ufual difcharges, fmall fhocks, *i. e.* of about one-twentieth of an inch, may be fent through the pelvis; fparks may be taken through the cloaths from the parts adjacent to the feat of the difeafe; and alfo the electric fluid may be tranfmitted by applying the metallic or wooden extremities of two directors to the hips, in contact with the cloaths; part of which may be removed in cafe they are too thick. Thofe various applications of Electricity fhould be regulated according to the conftitution of the patient. The number of fhocks may be about twelve or fourteen. The other applications may be continued for two or three minutes; repeating the operation every day. But either ftrong fhocks or a ftronger application of Electricity, than the patient can conveniently

bear,

bear, fhould be carefully avoided ; for by thofe means, fometimes more than a fufficient difcharge is occafioned, which is not eafily cured,   In cafes of uterine hæmorrhages, I don't know that the application of Electricity was ever beneficial, neither that it has been often tried.——Perhaps a very gentle electrization, as to keep the patient infulated and connected with the prime Conductor, whilft the electrical machine is in action, may be of fome benefit.

In refpect to *unnatural difcharges* and *fluxes* in general, it may be obferved, that fome difcharges are quite unnatural or adventitious, as the fiftula lacrymalis, and fome fpecies of the venereal difcafe ; but others are only increafed natural difcharges, fuch as the menfes, perfpiration, &c.   Now the power of Electricity, in general, has been found more beneficial for the firft, than for the fecond fort of difcharges, which are moftly increafed by it.

In the *venereal difeafe* electrization has been generally forbidden ; having moftly increafed the pains, and other fymptoms, rather

rather than diminifhed them. Indeed, confidering that any fort of ftimulus has been found hurtful to perfons afflicted with that diforder, it is no wonder that Electricity has produced fome bad effects, efpecially in the manner it was adminiftered fome time ago, *viz.* by giving ftrong fhocks. However, it has been lately obferved, that a very gentle application of Electricity, as drawing the fluid by means of a wooden or metal point, is peculiarly beneficial in various cafes of this kind, even when the difeafe has been of long ftanding. Having remarked above, that tumors, when juft beginning, are difperfed, and that unnatural difcharges are gradually fuppreffed by a judicious electrization; it is fuperfluous to defcribe particularly thofe ftates of the venereal difeafe in which Electricity may be applied; it is only neceffary to remind the operator to avoid any confiderable ftimulus in cafes of this fort.

The application of Electricity has been found alfo beneficial in other difeafes befides thofe mentioned above; but as the

facts

facts are not fufficiently numerous, fo as to afford the deduction of any general rules, I have not thought proper to take any particular notice of them; efpecially, becaufe the effects of Electricity on the human body, in various circumftances, have been already fufficiently confidered under general and comprehenfive heads.

We may laftly obferve, that in many cafes, the help of other remedies to be prefcribed by the gentlemen of the faculty, is required to affift the action of Electricity, which by itfelf would perhaps be ufelefs; and on the other hand, electrization may often be applied to affift the action of other remedies, as of fudorifics, ftrengthening medicines, &c.

" While I was writing fome of the
" above cafes," fays Mr. BECKET, " an
" obfervation or two occured to me, which,
" though perhaps of no great confequence,
" may not be amifs to mention, as every
" particular effect of Electricity feems to
" be worthy of notice.

" One

" One circumſtance attending ſome of
" the preceding cures, particularly that
" of the *paralytic*, related by Mr. JONES,
" was a freſh and copious diſcharge of the
" *bliſters*, which had been previouſly ap-
" plied to the patients.—This, I think,
" ſeems to be a pretty general conſequence
" of electrification; at leaſt, I have myſelf
" known many inſtances of it; particu-
" larly in one gentleman, whom I elec-
" trified for a paralytic complaint, and
" who had a bliſter applied to the back
" part of his neck. He informed me,
" that, in the night after his being elec-
" trified the preceding day, he found a
" much more copious diſcharge from the
" bliſter than at other times; though the
" operation was no more than his ſtand-
" ing, for about a quarter of an hour, on
" the inſulated ſtool, while ſparks were
" drawn from the ſide of his face. From
" hence it appears not impoſſible, that, in
" ſome caſes, bliſters may be attended with
" peculiar benefit, during a courſe of elec-
" trical treatment; in others, perhaps, it
" might be worth while to make uſe of
" Electricity, merely to obtain a favourable
" diſcharge from the bliſters."

VOL. II.   M   *Authentic*

Authentic phyſical Caſes, in which Electricity was adminiſtered.

C A S E I.

The particulars of the following caſe were communicated to me by Mr. PAR-TINGTON.

DANIEL WYSCOYL, aged thirty-ſix, of a ſtrong robuſt conſtitution, was ſent from the Weſtminſter Diſpenſary, in Gerard-ſtreet, to Mr. PARTINGTON, in order to be electrified for a violent inflammation in both his eyes. The account he gave of his diſorder, was the following :—Several dark objects of different ſhapes and ſizes, ſeemed at firſt to obſtruct his ſight. This was ſucceeded by an inflammation in both his eyes, which increaſed with ſuch rapidity, that in a week's time he was brought to the degree of blindneſs that afflicted him till he was electrified. He was im-mediately recommended to the Weſtmin-ſter Diſpenſary, where every poſſible at-tention was paid to his misfortune by

Mr.

Mr. FORD, the furgeon of that place; but the obftinacy of the diforder was fuch, that every endeavour made towards the relief of this poor man proved ufelefs.—— Blifters and leeches, befides the other ufual means, were applied without any efficacy whatever.

About two months after the commencement of the inflammation, Mr. FORD recommended him to Mr. PARTINGTON; who, on examining him, found that the eyelids could not be opened without the help of the fingers, and that when opened, the coats of the eye appeared of an uniform red colour. The fight of the right eye, which was the moft affected, was fo far impaired, that when it was turned towards a window, the eye-lids being forced open, he could perceive only a red glare of light like a ball of fire; but the reft of the room feemed to be equally dark, fo that he could not diftinguifh any object in it. With the left eye he could diftinguifh colours, and the fhapes of objects that were held to him, but in their fizes he was commonly miftaken. This diforder was accompanied with excruciating

M 2 pains,

pains, fhifting from one part to the other,
but principally infifting on his temples,
and fometimes darting to the back part of
his head, or to the centre of his eyes.

Mr. Partington began to electrify
him the 21ft of October, 1776 ; and three
days after the inflammation began vifibly to
abate, and in a fortnight's time it was quite
fubfided; but the pupil of the eye was fo
nearly clofed, that fcarce any of it could
be feen.   He continued to be electrified
every day for five weeks, and the pupil
gradually dilated, till he attained a degree
of fight fufficient to diftinguifh object
on the other fide of the way.   The pain
had now entirely left him, fo that he
omitted the ufe of Electricity, and did no
experience any farther inconvenience after
it.

This remarkable cure was effected by
throwing the ftream of electric fluid with
a metal, and with a wooden point. The
firft inftrument ufed, which was contrived
by the late Mr. Ferguson, confifted in a
pointed brafs wire, faftened by means o
a cork

a cork at the finaller end of a conical glafs, open at both ends, and paffing through the axis of this conical or funnel-like glafs, its point came within about half an inch of the larger aperture of the glafs. This inftrument being defigned to throw the electric fluid upon the eye, was to be fixed fo that the larger aperture of the glafs furrounding the eye kept its lids open, and the point of the wire was oppofite to the pupil, and about half or one inch from it. With this inftrument it was obferved that a fpark often proceeded from the point of the wire, which occafioned an infufferable pain; for which reafon Mr. PARTINGTON, who fpares no pains to advance this branch of phyfic, thought of improving this inftru-ment by fixing a wooden point upon the pointed wire, by which means the former inconvenience was entirely removed, and the ftream of electric fluid was rendered more efficacious, and more eafily manage-ble.

This, as far as I am informed, was the firft time that this moft excellent method of throwing the electric fluid, *viz.* with a wooden point, was ufed.

M 3   N. B. The

N. B. The directors described in the preceding pages, the principal of which were contrived by Mr. PARTINGTON, answer every required purpose, much better than the above-described inftrument of Mr. FERGUSON.

C A S E II.

The following cafe is related by Mr. LOVETT, in his *Electricity rendered ufeful:*

" Having obferved the great efficacy of the electrical æther, in foon relieving moft kinds of inflammations, I was inclined to think the fame falutary effects would appear when applied to the St. Anthony's Fire; but when a cafe of that fort offered, the inflammation was fo great, that at firft fight I almoft defpaired of fuccefs.

" About the middle of the day I made the firft trial, and before night the fwelling was much abated, and in a few days quite cured. .

" The operation was fimply drawing fparks with a finger, or an iron ftyle, while the

the perfon was electrified on the infulating
ftool."

C A S E III.

The following cafe is alfo related by
Mr. LOVETT:

" ANN THOMPSON, in Little Fifh-ftreet,
Worcefter, was troubled with a fiftula near
the inner corner of her eye, which broke
out, and healed, no lefs than feven times.
The laft time it healed, it continued well
for fome time; after which it began with
a fmall fwelling, and continued growing
larger, till it was as big as a filbert; when
fhe was advifed to try Electricity. After
the fwelling was electrified, it foon decreafed, till it was entirely difperfed; and
has continued well for more than two
years, without the leaft fymptom of any
return of the diforder. — The operation
was fimply drawing fparks from the part
affected."

C A S E IV.

The late Mr. FERGUSON being at Briſ-
tol, was ſeized with a violent ſore throat,
ſo that he could not ſwallow any thing.
Being willing to try the power of Electricity, Mr. ADLAM, of that city, performed the operation; which was merely
drawing ſparks from the throat. The
electrization was repeated half an hour
after, and was attended with ſo good. and
remarkable an effect, that in about one
hour's time Mr. FERGUSON could both eat
and drink without pain.

C A S E. V.

The following two caſes are related by
Mr. JOHN BIRCH, Surgeon.

" A young woman, at the age of twenty-
two, deſired my advice for a tumor on her
thigh, which followed an unhappy accident ſhe met with two years before. Her
caſe was attended with many complicated
ſymptoms, and, among them, a ſuppreſ-
ſion

fion of the menfes, which had lafted feven months.   I thought it right to relieve, if poffible, this fymptom, before I proceeded. to perform the operation, which was ne-ceffary for the tumor.

" For three fucceffive days I paffed fome electric fhocks through the region of the pelvis ; and on the fourth, fhe was attack-ed with a violent pain in her fide, which left her on applying the fhocks to that part. In about three hours it returned, and I was fent for.   I repeated the fhocks, and the pain again vanifhed.   I vifited her fix hours after, when the pain had begun to attack the fide.—I paffed a ftronger fhock, which removed it, and fhe flept well the whole night.—The next day, being the fifth, the menfes appeared, and flowed gently for three days ; but ceafing then, the pain of the fide returned, and was fo violent, that I was fent for in a hurry.—When I came to her, I found her in great agony ; but be-ing informed of the caufe, I begged to make trial of Electricity once more, which fhe readily confented to, as fhe had ex-perienced fuch inftantaneous relief before.
—On

—On its application, the pain ceafed.—A very fhort time after, the flux came on, and continued two days.—I attended her for feveral weeks after, upon the former account, and had the pleafure to fee her recover from all her complaints."

C A S E VI.

" I was fent for to a lady, who had been afflicted with painful ulcers on both her legs, for more than fifteen months. —They came after a lying-in, and had never healed. The legs were fwelled, but the ulcers had no malignant appearance.—She told me, that fince her laft mifcarriage, which was then more than ten months, fhe had never been regular.— She attributed the pain and fwelling of her legs to that caufe; and, upon enquiry, I found that fhe was fenfible of an endeavour of nature to relieve herfelf at regular periods, and that the pain fhe fuffered at thofe times was alleviated by a bloody difcharge from the ulcers. — I applied the proper dreffings and bandages to the parts, and waited the approach of that period.

In

In about ten days, a pain feized her back, and (he began to complain of her legs : I then electrified her ; and the next day fhe was taken out of order, and continued fo the whole week.—The ulcers mended from that time, and were healed in three weeks afterwards."

The reader may reft affured, that cafes of this fort are fo frequent, that perhaps Electricity may be confidered as a fure remedy for the arreft of thofe natural fluxes.

C A S E VII.

The following cafe is extracted from the LXVIIIth volume of the Philofophical Tranfactions.

A Cure of a mufcular Contraction by Electricity. By Mr. MILES PARTINGTON, *in a Letter to* WM. HENLY, F. R. S.

Charles-ftreet, Cavendifh-fquare, June 13, 1777.

"DEAR SIR,

" It is fome time fince, you informed
" me that you had mentioned to Sir JOHN
3   " PRINGLE,

" Pringle, Mifs Lingfield's cure by
" Electricity; that it excited his attention;
" and was his opinion, that the com-
" munication of it to the Royal Society
" would be deemed important and ufeful.
" I hope you will not blame my delay in
" the compliance with your requeft. I
" have waited for no other purpofe, than
" to obtain the lateft account of the per-
" manency of thofe good effects, which
" fhe had then but recently experienced,
" from our electrical experiments upon
" her. Of thefe advantages we have both
" had repeated confirmation; and I may
" now, I believe, with ftrict propriety,
" from the notes I made for my own fatis-
" faction, fubmit the following particulars
" of them to the infpection of whomfoever
" your judgment fhall direct, or to appro-
" priate them to any other purpofe you
" pleafe. As you were prefent when I firft
" waited on this unhappy young lady, you
" will recollect the condition in which we
" found her. Her head was drawn down
" over her right fhoulder; the back part
" of it was twifted fo far round, that her
" face turned obliquely towards the oppo-
" fite

" fite fide, by which deformity fhe was
" difabled from feeing her feet, or the fteps,
" as fhe came down ftairs. The *fterno-*
" *moftoideus* mufcle was in a ftate of con-
" traction and rigidity. She had no ma-
" terial pain on this fide of her neck; but,
" owing to the extreme tenfion of the
" teguments of the left fide, fhe had a pain
" continually, and often it was very vio-
" lent, particularly in fudden changes of
" the weather. Her pulfe was weak,
" quick, and irregular. She was fubject
" to a great irritability; had frequently a
" little fever, which came on of an even-
" ing, and left her before morning: her
" fpirits were generally exceedingly op-
" preffed; and at times fhe was flightly
" paralytic.

" She dated the origin of her diforder
" at fomething more than two years from
" that period. She was fuddenly feized,
" going out of a warm room into the cold
" air, with a pain upon the back of her
" head, which admitted of fmall abate-
" ment for fome months, contracting gra-
" dually the mufcles to the melancholy
" deformity

" deformity we then beheld; and not-
" withſtanding every prudent means had
" been uſed to ſubdue it, and ſhe ſtrictly
" adhered to every article preſcribed to her
" by the faculty, ſhe was ſenſible of little
" variation ſince, and that rather on the
" unfavourable ſide.

" I urged her to make a trial of Electri-
" city. She was willing while ſhe was in
" London to try the experiment; and
" though the weather was remarkably
" tempeſtuous, ſhe came to me the firſt
" tolerable day, and was electrified the firſt
" time, February 18, 1777.

" I placed her in an inſulated chair, and,
" connecting it by a chain to the prime
" Conductor of a large electrical machine,
" I drew ſtrong ſparks from the parts af-
" fected, for about four minutes, which
" brought on a very profuſe perſpiration
" (a circumſtance ſhe had been unaccuſ-
" tomed to) which ſeemed to relax the
" *moſtoideus* muſcle to a conſiderable degree;
" but as the ſparks gave her a good deal of
" pain, I deſiſted from drawing them, and
" only

" only fubjected her a few minutes longer
" to the admiffion of the fluid, which pafied
" off without interruption from the pores
" of her fkin and adjacent parts. The next
" time fhe came to me, was the 24th of
" the fame month. As fhe had been in the
" afternoon of the firft day's experiment a
" good deal difordered, I changed the mode
" of conducting the operation, and fat her in
" a common dining chair, while I dropped,
" for five minutes, by the means of a large
" difcharging rod with a glafs handle, very
" ftrong fparks upon the *moftoideus* mufcle,
" from its double origin at the *fternum* and
" *clavicula* to its infertion at the back of the
" head. She bore this better than before,
" and the fame good effect followed in a
" greater degree, and without any of the
" fubfequent inconveniences. I faw her
" the third time on the 27th. She affured
" me fhe had efcaped her feverifh fymp-
" toms of an evening, and that her fpi-
" rits were raifed by the profpect of getting
" well; that fince the laft time I electri-
" fied her, fhe had more freedom in the
" motion of her head, than fhe had ever
" experienced fince the firft attack of her
" diforder.

" diforder. I perfifted in electrifying her
" after the fame manner, March 3d, 5th,
" 6th, 7th, and 9th ; from each time fhe
" gained fome advantage, and her feverifh
" tendency and nervous irritability went
" off entirely.

" The weather now fetting in very un-
" favourable, and fearful of lofing the
" advantages we had happily reaped from
" our early efforts, I requefted the favour
" of you, as the next-door neighbour, to
" electrify her every evening while fhe
" was in town, and fhe might, if any al-
" teration took place, fee me occafionally.
" Fortunately for her, you accepted the
" propofal ; and to your judgment and
" caution in the conduct of it, for the next
" fortnight (three evenings only excepted)
" you brought about the happy event ; and
" have received her teftimony of gratitude,
" for relieving her from a condition, under
" which life could not be defirable, to a
" comfortable affociation with her family
" and friends.

" I am, &c."

" The

" The method I purfued was, to place the lady upon a ftool with glafs legs, and to draw ftrong fparks, for at leaft ten minutes, from the mufcles on both fides of her neck. Befides this, I generally gave her two fhocks, from a bottle, containing fifteen fquare inches of coated furface, fully charged, through her neck and one of her arms, croffing the neck in different directions. This treatment fhe fubmitted to with a proper refolution; and it gave me fincere pleafure to find it attended with the defired fuccefs.

W. HENLY."

C A S E VIII.

The following cafe is extracted from the LXIXth vol. of the Philofophical Tranf-actions.

An Account of a Cure of the St. Vitus's Dance, by Electricity. In a Letter from ANTHONY FOTHERGILL, *M. D. F. R. S. at* North-ampton, *to* W. HENLY, *F. R. S.*

Northampton, Oct. 28, 1778.

SIR,

Agreeable to my promife, I now pro-eed to give you fome account of a recent cure performed by Electricity, which will I think, afford you much pleafure.

Ann Agutter, a girl of ten years o age, of a pale emaciated habit, was admittec an out-patient at the Northampton Hofpi- tal on the 6th of June laft. From hei father's account it appeared (for fhe wa: fpeechlefs, and with difficulty fupportec from falling by two affiftants) that fhe hac for fix weeks laboured under violent con- vulfive motions, which affected the whole frame, from which fhe had very fhort in- termiffions, except during fleep; that the difeafe had not only impaired her memory and intellectual faculties, but of late hac deprived her of the ufe of fpeech.

Volatile and fetid medicines were now recommended, and the warm bath every other night, but with no better fuccefs, except that the nights, which had been reftlefs, became fomewhat more compofed. Blifters and anti-fpafmodics were directed, and particularly the flowers of zinc, which were continued till the beginning of July, but without the leaft abatement of the symptoms :

fymptoms ; when her father growing im-
patient of fruitlefs attendance at the hofpi-
tal, I recommended, as a *dernier refort,* a
trial of Electricity, under the management
of the Rev. Mr. UNDERWOOD, an ingenious electrician.   After this I heard no more
of her till the firft of Auguft, when her father came to inform me that his daughter
was well, and defired fhe might have her
difcharge.   To which, after expreffing my
doubts of the cure, I confented ; but fhould
not have been perfectly convinced of it, had
I not received afterwards a full confirmation
of it from Mr. UNDERWOOD, dated September 16; an extract from whofe letter I
will now give you in his own words.

" I have long expected the pleafure of
" feeing you, that I might inform you
" how I proceeded in the cure of the poor
" girl.   As the cafe was particular, I have
" been very minute, and wifh you may
" find fomething in it that may be ufeful
" to others.   If you think it proper, I beg
" you will ftate the cafe medically, and
" make it as public as you pleafe.

N 2   " July

" July 5. On the glaſs-footed ſtool for
" thirty minutes : ſparks were drawn from
" the arms, neck, and head, which cauſed
" a conſiderable perſpiration, and a raſh
" appeared in her forehead. She then
" received ſhocks through her hands, arms,
" breaſts, and back ; and from this time
" the ſymptoms abated, her arms begin-
" ning to recover their uſes *.

" July 13. On the glaſs-footed ſtool
" forty-five minutes : received ſtrong
" ſhocks through her legs and feet, which
" from that time began to recover their
" wonted uſes ; alſo four ſtrong ſhocks
" through the jaws, ſoon after which her
" ſpeech returned.

" July 23. On the glaſs-footed ſtool
" for the ſpace of one hour : ſparks were
" drawn from her arms, legs, head, and
" breaſt, which for the firſt time ſhe very
" ſenſibly felt ; alſo two ſhocks through
" the ſpine. She could now walk alone ;
" her countenance became more florid,
" and all her faculties ſeemed wonderfully

* The coated bottle held near a quart.

" ſtrengthened ;

" ftrengthened; and from this time fhe
" continued mending to a ftate of perfect
" health.

" Every time fhe was electrified pofi-
" tively, her pulfe quickened to a great
" degree, and an eruption, much like the
" itch, appeared in all her joints."

Thus far Mr. UNDERWOOD. To complete the hiftory of this fingular cafe, I
this day (Oct. 28) rode feveral miles, on
my return from the country, to vifit her;
and had the fatisfaction to find her in good
health, and the above account verified in
every particular; with this addition, that
at the beginning of the difeafe, fhe had
but flight twitchings, attended with running, ftaggering, and a variety of invo-
luntary gefticulations, which diftinguifh the
St. Vitus's Dance; and that thefe fymptoms were afterwards fucceeded by convul-
fions, which rendered it difficult for two
affiftants to keep her in bed, and which
foon deprived her of fpeech and the ufe of
her limbs. The eruptions which appeared
on the parts electrified foon receded,

N 3   without

without producing any return of the symptoms, and therefore could not be called
critical, but merely the effect of the electrical stimulus. Having given her parents
some general directions as to her regimen,
&c. I took my leave, with a strong injunc-
tion to make me acquainted, in case she
should happen to relapse. Before I conclude, it may not be improper to observe,
that some time ago, I was fortunate enough
to cure a boy who had long had the St.
Vitus's Dance (though in a much less degree) by Electricity. A violent convulsive
disease, somewhat similar to the above,
though, if I recollect right, not attended
with the *aphonia*, was successfully treated
in the same way by Dr. WATSON, and is
recorded in the Philosophical Transactions.
May we not then conclude, that these facts
alone, and more might perhaps be produced, are sufficient to intitle Electricity to a
distinguished place in the class of anti-spas-
modics ?

I am, &c.

C A S E IX.

The following is one of thofe cafes in which the ufe of Electricity was attended with bad fuccefs: this cafe was related by Dr. HART of Shrewfbury. See the Philo-fophical Tranfactions, VOL. XLVIII.

A girl, aged about fixteen, whofe right arm was paralytic, being electrified the fecond time, became intirely paralytic, and remained in that ftate for about a fortnight; then the fuperadded paralify was removed, by means of fome medicines; but the arm which was before paralytic, remained fo. It fhould be alfo added, that this arm was very much wafted in comparifon to the other. Notwithftanding the firft bad accident, it was refolved to make another trial of Electricity. But after ufing this treatment for three or four days, fhe became again univerfally paralytic, and even loft her voice, and could fwallow with difficulty. This fecond accident plainly fhewed

the

the bad effects of Electricity in that cafe, and the girl, although afterwards relieved of her additional paralify, remained in the fame ftate fhe was before the ufe of Electricity.

In this cafe, it is fufpected that Electricity was improperly managed; at that time it being ufual to give ftrong fhocks, which perhaps were pernicious in the above-mentioned cafe.

C A S E X.

The following cafe was performed under the direction of the learned Dr. Wm, Watson, F. R. S.

A girl belonging to the foundling-hofpital, aged about feven years, being firft feized with a diforder occafioned by the worms, was at laft, by an univerfal rigidity of the mufcles, reduced to fuch a ftate, that her body feemed rather dead than alive. After that other medicines had been ineffectually administered

adminiſtered for about one month, ſhe was at laſt electrified intermittedly for about two months, after which time ſhe was ſo far recovered, that ſhe could, without pain, exerciſe every muſcle of her body, and per-form every action as well as before ſhe had the diſtemper.

———————

The intelligent reader muſt have un-doubtedly remarked, that in ſome of the above caſes, the electrization adminiſtered was rather-ſtrong, and different from the general rules given previous to the narra-tion of the caſes. But-it muſt be obſerved, that ſome of the above caſes happened before the principal methods of electrify-ing, which are now uſed by the beſt prac-titioners, were introduced.—Perhaps, in ſi-milar caſes, the ſame ſalutary event might be produced by a more gentle electrization,

A Letter

A Letter from Mr. Miles Partington
to the Author.

Cavendifh Square, Auguft the 10th, 1781.

SIR,

As-the poffibility of a cure of a fiftula
lacrymalis by Electricity has been publicly
queftioned, I am very glad to comply with
your requeft, in ftating the fuccefsful treatment of a difeafe of this kind under my
own infpection.

Ann Woodward, between 20 and 30
years of age, was recommended to my care
by Mrs. Swift, of King Street, Bloomf-
bury, in whofe fervice fhe then lived —I
was told fhe had a fiftula lacrymalis, which
had refifted every attempt to relieve her.
I fhall defcribe the fituation fhe was in
when I firft faw her, and fhall add her own
account anterior to that period. There was
a very violent inflammation in the left eye,
attended with exceffive pain, and almoft
continual flux of tears down the cheek, the
fharpnefs of which had confiderably exco-
riated

riated the ſkin. A little prominence was
perceivable on the inner angle of the eye,
from which might be generally preſſed a
ſmall quantity of matter. The inflammation had been kept up with little abatement
for ſix months, during which time a cooling
regimen had been particularly enjoined her,
and was ſtrictly complied with. Since the
inflammation became violent, ſhe had conſtantly awoke in pain, which laſted till
twelve o'clock at noon, and was then tole-
rably eaſy the reſt of the day. She told me,
that ſhe had, as long as ſhe remembered,
been ſubject to a weak and watery eye, but
it had never given her much concern, till
ſhe received a cold in it upon her coming
to London. In every other reſpect ſhe was
remarkably healthy.

My firſt object was to relieve her of the
pain. For this purpoſe I conveyed the
electric fluid from a wooden point, which
generally at the time blunted the accute-
neſs of the ſenſation ; but which, if by
bringing it near to her eye the fluid was
concentrated, it always increaſed it to a
great degree, though this went off by
being

being left to itfelf, or repeating the milder method. I continued in this courfe of treatment for three weeks, when the inflammation was nearly fubfided, and the pain entirely gone away. Still perceiving the matter to ooze from the inner angle, I ventured to pafs a fingle electric fhock down the duct of the nofe, which I effected by placing one of the directors upon the lacrymal fac, and the other up the noftril, for the convenience of a more immediate local conveyance. This gave her much pain after the operation, and it remained all the reft of the day. I found in the morning that fhe could hardly bear me to prefs upon the fac, and very little matter came from it. I then paffed four fhocks, in the fmalleft degree I could convey them, and to be felt, which were not attended with fo much pain. At bed-time there came on a great throbbing in the part, and in the courfe of the night a large quantity of matter burft down the noftril, when fhe became immediately eafy. Some matter continued to come away for about four days, and fhe appeared to be perfectly well. Her eye has been fince in a ftronger ftate than it ever was before.

If

If this should still want additional
strength, I am willing to give you farther
instances of relief in this disease by Electricity, though they have not been so effectually cured.

I shall now add a few observations, which
may perhaps merit your notice.   In order
to increase the power of electrization, I
have added a very large coated jar to the
prime Conductor of my electrical machine.
This jar is placed upon a mahogany stand,
so that the knob of it touches the Conductor.   The insulated chair is also in contact
with the said Conductor, and the patient is
seated in it as usual.   With this disposition
of the apparatus, the wheel of the machine
is first turned a few times round; then I
apply the metallic rod to the patient, in order to draw the sparks through the cloaths,
or the stream of electric fluid by a wood-
en point, if the disease seem to require
such treatment; and I find, that by this
means the effects of the electrization are
considerably increased, the pungency of the
sparks is felt much deeper into the electrified part of the body; the heat occasioned
by

by it is alfo greater, and therefore feems
to be more efficacious for internal complaints. Add to this, that the ufual inconvenience of the diffipation of the electric
fluid into the air is confiderably prevented.

This difpofition of the apparatus does
alfo anfwer another purpofe, which is that
of electrifying a patient, without having
any affiftant in the room; for after the jar
is charged, the turning of the wheel may be
difcontinued, and the patient may continue
under the electrization. No fhock is to be
feared from this apparatus; for, fince the
wooden ftand upon which the jar is fixed is
a bad Conductor, the difcharge can only be
made gradually.

I have made another addition to my directors, or rather have contrived a new director, by which an electrified ftream of
water or other fluid is thrown upon any
part of the body. This director confifts
of a fmall glafs tube about 3 inches long,
and a quarter of an inch in diameter, one
extremity of which is drawn to a very fine
point, fuch as the water can hardly pafs
through.

through. This tube I faften to an infulating handle, which being made like a pair
of pincers, holds it by the middle ; I then
pour a little water or other fluid into the
tube, and by means of a wire connect it
with the prime Conductor. When the
machine is in motion, the ftream of water,
which the electric fluid forces out of the
tube, is very much fubdivided, and, when
directed upon the face, or any naked part
of the body, gives an agreeable fenfation,
which not only proves very refrefhing to
patients, but in cafes of great irritation of the
nerves is often attended with permanent relief. The fhort experience I have had of
thefe directors does not enable me to determine how far they may be ufeful ; but in
feveral inftances they have afforded confiderable relief, when other modes of electrization proved ufelefs.

In the courfe of my practice, I have obferved a very remarkable effect of Electricity upon the human body, which is, that
it removes coftivenefs in thofe perfons that
are electrified, efpecially along the courfe
of the alimentary canal. I muft obferve,

6   that

that it does by no means increafe the eva-
cuations of ordinary good habits of body,
but only reinftates the ufual difcharge in cafe
of coftivenefs. This effect feems to take
place becaufe the electrization gives vigour
and energy to the fibres of the debilitated
inteftines, in the fame manner as it re-
ftores the loft motion of more external
mufcles.

I am,

Dear Sir,

Yours, &c.

Miles Partington.

A Letter

A Letter from Dr. JAMES LIND *to the Author.*

Windfor, June the 17*th,* 1784.

DEAR SIR,

I here fend you the account of a remarkable cafe relieved by Electricity whilft I was at Bombay : if you think it worth publifhing, I beg you will infert it in the next publication of your Medical Electricity.

" The wife of an officer of the artillery at Bombay, during the laft months of her pregnancy, gradually loft the ufe of her lower limbs, as if it had been a paralyfis, occafioned by the preffure of the fœtus upon the nerves which go to thofe extremities. Her pregnant ftate did not allow the application of any remedy for the relief of the diforder. She came to her full time, was fafely delivered, and though fhe foon recovered in every other refpect, yet, contrary to every expectation, the paralyfis ftill remained; nor could it be in the leaft reliev-

VOL. II.   O   ed

ed by any of the various medicines, which were adminiftered for about feven months, which had elapfed from the time of her delivery, and until I firft vifited her, which was in June 1780.

" Upon enquiry finding that, of the many remedies likely to afford any relief in fuch a cafe, Electricity alone had not yet been tried, I immediately recommended the application of it. But the difficulty was to excite an electrical machine in an atmofphere fo exceffively moift as that of Bombay was at that time, the rainy feafon being already fet in ; however, the hufband of the patient, being the fuperintendant of the Military Laboratory of that place, propofed to try whether an electrical machine could be made to act in a heated room ; fuch as they ufed for the drying of gunpowder. Accordingly a fmall ftucco'd room in his houfe was heated by means of burning charcoal, then the doors and windows were thrown open, and an electrical machine being brought in, was found to act very powerfully. Things being thus prepared, my patient began to be electrified, firft by

3   giving

giving fparks to her legs and thighs, and
afterwards by paffing about twenty very
fmall fhocks up one leg and down the other.
The effect was really furprifing; for, after
the firft electrization, fhe was fo far relieved
as to be able to walk up fome fteps without
any help, which fhe had not been able to do
for many months before.   By the fecond
days electrization, which was performed exactly as in the preceding, fhe was enabled
to walk out, and vifit feveral of her friends
in the neighbourhood.   The third day's
electrization compleated the cure, and fhe
went about with all the eafinefs and alacrity
in the world.   I afterwards received a letter
from her hufband, dated May the 29th,
1781, informing me of her continuing in
perfect health."

Faithfully yours.

O 2   APPENDIX.

APPENDIX.

N° I.

OF THE VINDICATING ELECTRICITY.

1. AB, *ab*, fig. 6, of Plate IV. reprefents a plate of glafs, coated on both fides with the two metallic coatings CD, *ca*, which are not ftuck to the glafs plate, but are only laid upon it.

From the upper coating CD, three filk threads proceed, which are united at their top H, by which the faid coating may be removed from the plate in an infulated manner, and may be prefented to an electrified electrometer, as reprefented in fig. 7, in order to examine its Electricity. F G is a glafs ftand, which infulates and fupports the plate, &c.

2. Let

2 Let the plate A B, *ab*, be charged in the common manner, by means of an electrical machine, fo that its furface A B may acquire one kind of Electricity, (which may be called K) and the oppofite furface *a b* may acquire ·the contrary Electricity, (which we fhall call L). Then, if the coating C D be removed from the plate, and be prefented to an electrified electrometer, as reprefented in fig. 2, it will be found poffeffed of the Electricity K, *viz.* of the fame kind with that which was communicated to the furface A B of the glafs plate; from whence it is deduced, that the furface A B has imparted fome of its Electricity to the coating. Now, this difpofition of the charged plate to give part of its Electricity to the coating, is what the learned F. BECCARIA nominates the *Negative vindicating Electricity.*

3. If the coating be again and again alternately laid upon the plate and removed, its Electricity K will be found to decreafe gradually, till after a number of times (which is greater or lefs, according as the edges of the plate infulate more or lefs. exactly)

actly) the coating will not appear at all electrified. This state is called *the limit of the two contrary Electricities*; for if now the above-mentioned operation of coating and uncoating the plate be continued, the coating will be found poffeffed of the contrary Electricity, *viz.* the Electricity L. This Electricity L of the coating is weak on its firft appearance, but it gradually grows ftronger and ftronger till a certain degree; then infenfibly decreafes, and continues decreafing until the glafs plate has entirely loft every fign of Electricity.

By this change of Electricity in the coating, it is deduced, that the furface A B of the glafs plate changes property; and whereas at firft it was difpofed to part with its Electricity, now (*viz.* beyond *the limit of the two contrary Electricities*) it feems to *vindicate* its own property, that is, to take from the coating fome Electricity of the fame kind with that of which it was charged: hence this difpofition was by F. Beccaria called the *Pofitive vindicating Electricity.*

4. This

4. This pofitive vindicating Electricity never changes, though the coating be touched every time it is removed. It appears ftronger, and lafts a very confiderable time after the plate has been difcharged; which is a very furprifing property of glafs, and probably of all good and folid electrics.

5. If, foon after the difcharge of the plate, the coating be alternately taken from the plate, and replaced, but with the following law, *viz.* that when the coating is upon the plate, both coatings be touched at the fame time, and when the coating is off, this be either touched or not; then the furface A B of the plate, on being uncoated every time, takes a quantity of Electricity, which it alternately lofes every time it is coated *.

6. On removing the coating in a dark room, a flafh of light appears between it and the glafs, which is ftill more confpi-

* This may be proved by touching an infulated electrometer with the coating, when this is ftanding upon the plate, and when feparated from it.

O 4   cuous,

cuous, if the coating be removed by the fingers being applied immediately to it, *viz.* not in an infulated manner; becaufe, when the coating is not infulated, the glafs plate can give to, or receive from it, more of the electric fluid, and that more freely, than otherwife.

7. It is obfervable, that in the negative vindicating Electricity, the glafs lofes a greater or lefs portion of Electricity, in an inverfe proportion of the charge given to the plate, *viz.* the part loft is greater when the charge has been the weaker; for in the pofitive vindicating Electricity, the force of receiving Electricity is the ftronger, when the charge has been ftronger, and contrarywife.

8. If, after every time that the coating CD is removed, the atmofpheres E, *e*, that is, the air contiguous to the furface of the glafs plate, be examined, they will be found electrified as in the following table, *viz.* the threads of an electrometer, brought within one or two inches, or more, of the furfaces A B, *ab*, will diverge with Elec-

tricities

tricities contrary to thofe expreffed in the table.

During the time of the negative vindicating Electricity	the air E, if the plate has been charged	moderately high - - very high -	moderately L o moderately K
	the air *e* is electrified L.		

During the time of the pofitive vindicating Electricity	the air E The air *e*	are electrified L.

9. Although we are not acquainted with the caufe of vindicating Electricity, any farther than to confider it as a difpofition or property of charged glafs, yet the phenomena of the Electricities of the air, contiguous to the furfaces of the plate, feem to be a proper confequence of Dr. FRANKLIN's theory of Electricity, and are accountable by it; for it is a well-known principle of that theory, that when one fide of a coated electric, fit to receive a charge, acquires a greater quantity of Electricity than the oppofite fide can acquire of the contrary Electricity, then both fides of that electric appear poffeffed of the fame kind of Electricity, namely, of that communicated

municated to the firft-mentioned fide.
Now, when in the negative vindicating
Electricity, the furface A B of the glafs
plate gives part of its Electricity to the
coating, then the other fide *ab*, being more
electrified L, than the fide A B is electrified K, it is plain that, according to the faid
principle, both fides muft appear electrified
L.   But in the pofitive vindicating Electricity, the fide A B of the glafs plate receives
fome Electricity of the kind K from the
coating, therefore both fides muft affect the
air with the Electricity K.

10. There remains only to be explained
the reafon why, when the plate has received a high charge, the air E, during the
time of the negative vindicating Electricity, appears electrified K, whilft the air *e* is
electrified L.   In order to render this explanation more intelligible, let us fuppofe
the glafs plate to have been electrified pofitively on the fide A B; then, in the negative vindicating Electricity, the furface
A B, on being uncoated, lofes a part of its
Electricity, which is fo much the greater
as the charge has been the lefs (§ 7.);
therefore,

therefore, when the charge has been moderate, A B lofes a greater portion of electric fluid, than that with which the air *e* can fupply the furface *ab*; hence the furface *ab* will remain more negatively electrified than the furface A B is pofitively; confequently, according to the above-mentioned principle, (§ 9.) both the atmofpheres E, *e*, muft appear in a negative ftate when the charge has been a certain degree higher; then the furface A B, on being uncoated, lofes juft fo much of the electric fluid as the air *e* can give to *ab*, therefore the air will not appear electrified. But when the charge has been very high, A B lofes a lefs portion of electric fluid than the air *e* can give to *ab*; therefore *e*, by having given fome of its natural electric fluid to *ab*, will appear negative, and E will appear pofitive in a fmall degree. If the plate be fuppofed to have been charged negatively on the fide A B, the explanation of the phenomena is the fame, changing only the name of *pofitive* Electricity into *negative*, &c.

11. This property of charged glafs, called vindicating Electricity, is obfervable

also

alfo when two glafs plates, laid one over
the other, and coated on their outward
furfaces only, are charged jointly like one
plate.   Suppofe A B, C D, fig. 8, to reprefent the two plates charged together,
viz. by having prefented the coating F to
the prime Conductor, and having at the
fame time connected the coating G with
the ground, in which ftate, it is eafy to
conceive, that the upper furface of the plate
A B would be pofitive, its under furface
would be negative, the upper furface of the
plate C D, *viz.* the furface contiguous to
the plate A B, would be pofitive, and its
oppofite furface G would be negative.
Now if thefe plates, after having been
charged, be alternately feparated and joined, without ever touching their coatings,
it is plain that their furfaces, contiguous to
one another, whenever the plates are feparated, will uncoat each other, confequently
the phenomena of vindicating Electricity
will take place, that is, each of the infide
or naked furfaces, when the plates are firft
feparated, will lofe part of its Electricity.
This loft Electricity gradually decreafes
till it vanifhes, after which period, each of
the

the faid furfaces will gradually recover part
of its loft Electricity, &c.

12. By the principle noticed above,
(§ 9.) when one furface of either plate has
acquired a quantity of one kind of Electricity more than the oppofite furface has acquired of the other, then both furfaces of
that plate muft appear poffeffed of that and
the fame kind of Electricity; hence it follows, that when the plates A B, C D, are
at firft feparated for a certain number of
times, *i. e.* during the negative vindicating
Electricity, the plate A B muft appear pofitive on both fides, and the plate C D negative on both fides; but after the *limit of
the two contrary Electricities,* when the pofitive vindicating Electricity has taken place,
then the plate A B will appear negative on
both fides, and the plate C D pofitive on
both fides.

13. The adhefion of the plates to one
another keeps pace with the vindicating
Electricity; fo that it is very ftrong at firft,
but gradually decreafes with the negative
vindicating Electricity, till it becomes infenfible;

fenfible; but after the limit of the two contrary Electricities it appears again, and then it increafes and decreafes with the pofitive vindicating Electricity.

14. Every other particular relating to the phenomena of vindicating Electricity, exhibited with one plate, does alfo take place in the experiment with two plates; except the phenomena confidered above (§ 10.), which the two plates cannot exhibit, on account that they are not capable of receiving a very high charge, as a fingle plate is; which high charge is abfolutely neceffary to produce that appearance.

N° II.

N°. II.

OBSERVATIONS UPON THE CONDUCTORS OF LIGHTNING.

SINCE the publication of the First Edition of this Treatise, it happened that a house belonging to the Board of Ordnance, at Purfleet, was struck by lightning, though furnished with a Conductor. This accident excited anew the controversy relating to the construction of Conductors to secure houses from the effects of lightning, especially relating to their termination; and various experiments were made by very able Electricians, in order to decide the controverted point. Mr. B. WILSON and Mr. ED. NAIRNE, both members of the Royal Society, were the principal actors in this experimental investigation; the former giving constantly the preference to short Conductors terminating in a ball, and the latter preferring long Conductors acutely pointed; which has been and is the prevailing opinion with almost all the principal

cipal

cipal Electricians in England, as well as abroad; beginning from the firft inventor of thofe Conductors (the great philofopher Doctor FRANKLIN). Mr. WILSON, the principal and almoft the fole opponent of the Franklinian conftruction of Conductors, exhibited fome ingenious electrical experiments upon a large fcale; which threw great light upon the property of points in refpect to Electricity, and feem at firft fight to decide the queftion in favour of knobbed Conductors, as will appear by the following pages. On the other fide, Mr. NAIRNE performed fome well-imagined experiments, which feem to eftablifh, beyond any doubt, the prevailing opinion in favour of pointed Conductors; which was afterwards ftill more confirmed and elucidated by Lord MAHON, in his learned work, entitled, *Principles of Electricity,* and publifhed in the year 1779.

Notwithftanding thofe various difquifitions, and new obfervations, it feems that what I had previoufly written, relating to the conftruction of Conductors, did not require any material alteration; hence I thought proper to reprint, in this Third Edition,

Edition, exactly that which had been advanced relating to Conductors of lightning in my firſt publication; and have reſerved for this diſſertation, to make mention of thoſe additional remarks, which, in the preſent more advanced ſtate of the ſcience, are neceſſary to be noticed.   Mr. WILSON, Mr. NAIRNE, and others, muſt excuſe my giving a very ſhort and ſummary account of their valuable experiments, ſince the limits of this work are by far too ſhort to admit a full account of their obſervations. I ſhall juſt mention the principal facts aſcertained, ſo as to make evident the deduction which naturally follows them, and which is neceſſary to eſtabliſh the preſent improved theory of Conductors; omitting thoſe remarks which ſeem to be not eſſential, and even diſpenſing with the particular deſcription of the various apparatuſes uſed for this purpoſe.   The *Board-houſe*, ſtruck by lightning on the 12th of May 1777, had a pointed iron ſpike, which projected ten feet above the middlemoſt and higheſt part of its roof, and which had a continued metallic communication to the ground.   The lightning did no other

VOL. II.   P   damage

damage to this houfe, than to throw down a
few ftones from one of its upper corners,
which were next to two iron cramps;
which cramps had not a metallic communication with the Conductor. This corner was forty-fix feet diftant from the top
of the Conductor. In this cafe it is difficult to fay, whether the lightning ftruck
the Conductor firft, or the corner of the
houfe juft over the above-mentioned iron
cramps, and from thence, being paffed to
the metal which communicated with the
Conductor, was conducted to the ground,
without caufing any farther damage to the
houfe. In the firft cafe, the point of the
Conductor would, in all probabitity, have
been fufed, as is generally the cafe; but no
mark of fufion could be perceived on it.
The fecond cafe feems not very likely to
have happened, fince the Conductor was
far above the corner that was ftruck, and
fince the lightning is known to ftrike the
moft elevated objects, and thofe which have
a better communication with the earth, in
preference to any other *. In any cafe,
the

* Mr. NAIRNE concludes the account of his above-
mentioned experiments with the following paffage:

" I muft

the only useful remark that can be drawn
from this accident is, that all the metallic
parts, that are in a building, should be connected with the Conductor, otherwise it is
not unlikely, but that, either by a direct
stroke, or by a lateral explosion, the house
may suffer some damage from lightning.
Mr. WILSON, who had formerly declared
his opinion against elevated and pointed
Conductors, derived from this accident a
strong proof apparently in favour of his
theory,

" I must beg to intrude a little more on your time,
" to remark on that part of Mr. WILSON's paper,
" where, from his experiments, he seems to conclude,
" that the lightning at Purfleet first struck on the
" point of the rod of the Conductor, and then, by a
" lateral part of that stroke, struck the cramp on the
" coping-stone. I believe, if he had examined the
" situation of the stone, and the place where the
" cramp was struck, he would have found, that, if the
" lightning had struck on the point of the Conductor,
" that, to have produced that effect on the stone, it
" must, after it had struck on the point, and passed
" down a quantity, have struck from the metal up into
" the air, then down again on the cramp, and then
" again to the metal it had left; for the small dent or
" hollow made by the lightning was on the upper sur-
" face of the stone, and yet the metallic communication
" to the earth continued from the point under the stone

" which

theory, and foon began to perform feveral experiments, in which the natural
accident was imitated by art *. Mr.
Wilson's great apparatus was fixed in the
Pantheon, Oxford-ftreet. It confifted principally of the following parts :—A cylinder,
16 inches in diameter, and 155 feet long,
covered with tin-foil, formed the prime
Conductor, which was charged by a good
electrical machine of the common fort.
This vaft Conductor was fufpended by various filk ftrings, at a convenient diftance

" which was ftruck. It appears more probable to me,
" from the trifling damage it did, that the charged
" cloud had paffed over the pointed Conductor, and
" had been exhaufted of a great part of its Electricity
" in paffing; and that, after it had paffed, it was at-
" tracted down lower by a ridge of hills that was be-
" yond, and that the cloud being out of the influence
" of the point to prevent its ftriking, the end of the
" cloud might ftrike at an angle in the cramp, and fo
" to the metallic part of the Conductor, which was
" only about 7 inches below. I fhall conclude with
" obferving, that Mr. Henly and myfelf had the
" pointed rod of the Conductor at Purfleet taken down
" to examine the point; but we found no appearance
" on it that fhewed that it had been ftruck."

* The account of thofe experiments is inferted in
the 68th vol. of the Phil. Tranf.

from

from the floor and walls. To this Conductor. there was added a wire 4800 feet
long, which was also supported by silk
strings, and, on account of its prodigious
length, was bent in a great many places,
going backwards and forwards in various
directions. Sometimes this wire was separated from the great cylinder; but, for the
generality of experiments, they were made
to communicate together, so as to form one
extensive Conductor.—By means of a proper frame and machinery, a wooden model
of the above-mentioned house, that was
struck by lightning, was made to pass with
different required degrees of velocity under
and across one end of the great cylinder.
This model was furnished with a Conductor at top, which was sometimes terminated
in a point, and at other times ended in a
ball, but which was made to communicate
with the earth by as good a metallic communication as can be desired by the most
scrupulous experimenter. Whilst the model of the house was kept at a proper
distance from the great cylinder, and after a
known number of turns of the wheel, the
model being let go, passed under the end of

the

the cylinder with any required velocity, and at the moment that it paſſed imme- diately under the charged cylinder, it receiv- ed the ſtroke from it upon its Conductor, &c. Now the principal experiments which Mr. WILSON made with this appa- ratus, are briefly expreſſed in the following articles :

1. When the model was furniſhed with a ſharp Conductor, on being made to paſs under the charged cylinder, it drew off a good deal of the electric fluid from it in a ſilent manner, and that abſorption, as it may be ſaid, began long before the model came quite under the cylinder, as was ſhewn by the appearance of light, &c. The charge which afterwards remained in the cylinder was inconſiderable : all which ſhews the tendency that a point has to draw the electric fluid ſilently, and from a diſtance.

2. If, inſtead of a point, the model was furniſhed with a knobbed Conductor of the ſame length with the pointed one, the quantity of electric fluid drawn off in re- peating

peating the above experiment was not fo-
confiderable as before, and it was almoft
nothing, if the knobbed Conductor was
much fhorter than the pointed one, as could
have been expected.

3. When the model with the fharp
Conductor was made to pafs under the cylinder fully charged, the pointed Conductor
was generally ftruck with a full and ftrong
explofion; whereas, if the Conductor was
terminated in a ball (though of an equal
length with the pointed Conductor, and
confequently equally diftant from the cylinder when the model was juft under it)
it was not ftruck, the cylinder remaining
charged, &c.

4. " The weight which moved the mo-
" del in the preceding experiment, was
" gradually reduced till it was nearly ba-
" lanced by the friction; and when the
" motion was rendered fo flow as feven feet
" feven inches in feven feconds, it was very
" little accelerated; and in this ftate the
" great cylinder being charged, the model
" was fuffered to pafs; and, though the
P 4   " velocity

" velocity was lefs than three quarters of a
" mile in an hour, the point was ftruck."

The intelligent reader may remark, that
by the above-mentioned obfervations, the
already-known properties of points were
clearly fhewn : that the very reafons before
advanced by various philofophers, for preferring pointed Conductors of lightning to
fuch as are blunted, were confirmed ; but,
however, that one remarkable obfervation,
which could have been hardly believed
before, was eftablifhed, *viz.* that a point
in thofe circumftances could receive fo full
a ftroke, as we have mentioned above. Indeed this is a very good obfervation of Mr.
WILSON, and a moft plaufible argument
in favour of his opinion relating to the conftruction of the Conductors of lightning.
Hence he expatiates upon the danger of
thofe houfes, which are furnifhed with elevated and pointed Conductors; and abfolutely recommends knobbed Conductors,
which are either even with, or rather a little fhorter, than the top of a building ; fince,
fays he, the pointed Conductors invite the
lightning. It is almoft fuperfluous to remark,
that

that it is for that very property, that elevated and pointed Conductors are preferable to knobbed ones, *viz.* becaufe they attract the matter of lightning from a greater diftance : hence they may defend a greater extent of building ; hence, by drawing a confiderable quantity of electric fluid from the clouds, when at a diftance, fo as to leffen their charge, they may thus actually avoid a full explofion ; the intention of putting a Conductor to a houfe, being to defend the houfe from the bad effects of lightning, and not the Conductor itfelf from attracting the matter of, or being ftruck by, the lightning.—When the Conductor of a houfe is ftruck, the only damage it can receive in that cafe, it feems, can only arife from a lateral explofion between the Conductor and other pieces of metal, or other very good Conductors, which are contained in the houfe, and are not properly connected with the Conductor.   For this reafon, it is proper to connect with the Conductor, by a metallic communication, all the pieces of metal, and indeed other good Conductors, as a ciftern of water, &c. that are in a houfe, efpecially thofe which are near the outfide,

and

and the top of it. This, however, is almoſt
impoſſible to be done perfectly; but then
thoſe lateral effects are not of ſo very dangerous a nature, excepting indeed when
the building contains ſubſtances of a very
combuſtible quality, as gun-powder, &c. in
which caſe, the moſt ſcrupulous attention
is hardly ſufficient to prevent any bad accident. Notwithſtanding the obviouſneſs of
the above-mentioned remarks, Mr. Wilson's experiments determined ſeveral perſons in favour of the knobbed Conductors:
hence Mr. Nairne undertook to oppoſe
Mr. Wilson's theory with a clear and
convincing ſet of experiments. The prime
Conductor uſed by Mr. Nairne was one
foot in diameter, and ſix feet long, conſe-
quently by far ſmaller than that uſed by
Mr. Wilson. The other apparatus was
different, and the electrical machine was
much more powerful than that uſed by
Mr. Wilson *. But with this apparatus
he amply ſhows the properties of points and
knobs, with reſpect to Electricity; and, different from Mr. Wilson, ſhows, that in

* See Mr. Nairne's account of thoſe Experiments,
&c. in the 68th vol. of the Phil. Tranſ.

many

many experiments, and under various cir-
cumſtances, a knob is ſtruck in preference
to a point.   This diverſity of appearances,
and reſults of experiments, in caſes ſeemingly alike, will at firſt ſight be thought a
contradiction; but, when duly conſidered,
the different phenomena may be all recon-
ciled to a few very ſimple and natural
propoſitions.   A point draws the electric fluid from an electrified body, from a
much greater diſtance than a blunt-termi-
nated body; but the quantity muſt be limited, and ſubject to be altered by many
cauſes; ſuch are, the degree of condenſation of the electric fluid upon the electrified
body; the time given to the point; the
acuteneſs of the ſaid point; its free or encumbered ſituation; its perfect or imperfect communication with the earth, &c.
Thus, if the pointed body be not made to
communicate with the earth, but the communication be interrupted by a ſhort interval, then, on preſenting a body ſufficiently
charged with electric fluid, to the ſaid point,
a full ſpark will go from the former to the
latter, becauſe the point, on account of the
interruption of communication, cannot diſcharge

charge to the earth, or other fit body, the electric fluid by little and little, confequently cannot receive it filently from the electrified body to which it is expofed, but can attract only that quantity which can leap between the interruption, and which paffes in the form of a fpark. Thus alfo, if the point is fuddenly brought within a fufficient diftance of a ftrongly-electrified body, it will receive a fpark, becaufe a fufficient time was not allowed it to draw off the electric fluid filently, as would happen if the point was gradually brought near. After the fame manner, the ingenious reader may imagine a great multiplicity of cafes, in which a point may be ftruck or not ftruck by an electrified body; hence is derived the apparent contradiction between the experiments of Mr. HENLY, Mr. WILSON, Mr. NAIRNE, and various others: but it always remains true, that a pointed or fharp-edged body draws the electric fluid from a greater diftance than thofe bodies which are blunt or more obtufe, all the other circumftances being the fame in both cafes.

Lord

Lord MAHON, in his above-mentioned work, fhews a property of electric atmo-fpheres, which he calls the *returning ſtroke,* and which he is of opinion that can occa-fion great damage in fome cafes. This is that quantity of electric fluid, which, by the vicinity of a cloud highly electrified, is driven away from certain bodies, and which fuddenly returns to thofe bodies when the cloud happens to difcharge its Electricity by a ſtroke of lightning to any other body; for then the atmofphere which kept off the electric fluid of the bo-dies that were within its action, ceafes at once. Our noble author fhews experimen-tally, that fuch a returning ſtroke may oc-cafion great damage, even at a good diſtance from the place where the ſtroke of light-ning happens, and that its effects may be very confiderable; all which confirms the already-made obfervation, *viz.* that all the beſt and largeſt conducting bodies that are in a houfe, fhould be connected with the Conductor of lightning. Without detain-ing my reader with further difquifitions, I fhall only take notice of the moſt notable particulars that fhould be kept in view,

and

and fhould be confidered as the leading
principles, in erecting Conductors for the
lightning; and fhall then conclude with
the .*requifites neceffary* for the proper conftruction of Conductors, which requifites I
fhall tranfcribe exactly from Lord MA-
HON's work.

A pointed body attracts the matter of
lightning more or.lefs eafily, according as
it is more or lefs acute; according as it is
lefs or more encumbered by furrounding
bodies; or as it is more or lefs elevated,
and as its communication with the earth,
by means of conducting bodies, is more or
lefs perfect.   The clouds are bad Conductors, and in general the electric fluid is not
highly concentrated on them in proportion
to their extent; they are often feparated
from each other; they may come towards
a Conductor or a houfe in every direction;
and move with various velocities at different
times.   The pointed Conductor that is
erected in order to tranfmit the lightning
with fafety to the ground, can only defend
a limited extent of building, the quantity
of which varies according to many circumftances.

ftances. In cafes of powder-magazines, and fuch-like very combuftible fubftances, great damage may be occafioned by a lateral explofion or returning ftroke, if the Conductor happened to be ftruck; which may be only avoided by making a proper communication between all the fubftances of a good conducting nature that are in the building. It has been obferved that fometimes a ftroke of lightning is branched, and ftrikes feveral objects at the fame time.

In confequence of all thofe obfervations, and various others mentioned in the courfe of this book, it feems proper to have in view the following inftructions, in erecting Conductors for the lightning.

1. " That the *rods* be made of fuch fubftances as are, in their nature, the beft *Conductors* of Electricity."

2. " That the rods be *uninterrupted,* and perfectly *continuous.*"

3. " That they be of a *fufficient thicknefs.*"

4. " That they be perfectly *connected* with the *common-ftock.*"

5. " That the upper extremity of the rods be as *acutely pointed* as poffible."

6

6. " That

6. " That it be *very finely tapered.*"

7. " That it be *prominent.*"

8. " That each rod be carried, in the *shortest convenient direction,* from the point at its upper end, to the *common-stock.*"

9. " That there be neither *large* nor *prominent* bodies of metal upon the top of the building proposed to be secured, but such as are *connected with the Conductor* by some proper metallic communication."

10. " That there be a sufficient *number* of high and pointed rods :" And,

11. " That every part of the rods be very *substantially* erected."

N° III.

N° III.

EXTRACT OF A LETTER FROM MR. AR-
DEN, LECTURER IN NATURAL PHILO-
SOPHY, DATED SEPTEMBER 25, 1772.

" ABOUT fourteen or fifteen years
ago, in the prefence of WM.
CONSTABLE, Efq; at his feat at Burton
Conftable, in Holdernefs, I made the fol-
owing experiments :

" I placed a large coated jar, that would
hold three or four gallons, directly under
he prime Conductor of a very good elec-
rical machine. The prime Conductor
was at leaft eight or ten inches above the
op of the jar, and the communication was
made by a brafs wire, bent at one end over
he prime Conductor, and the other end
paffed through a fmall glafs tube (con-
rived by Mr. CONSTABLE to prevent the
electric matter from eafily flying off) was
fufpended in the middle of the jar, and had
fmall piece of brafs chain faftened to it,
hat refted on the bottom of the jar.

VOL. II.   Q   " I then

" I then began to turn the wheel, and
after turning about 100 or 150 times, a
low in the jar as·I could fee for the coating
I perceived a ball of fire, much refem
bling a red-hot iron bullet, and full thre
quarters of an inch in diameter, turnin
round upon its axis, and afcending up th
glafs tube that contained the brafs wire
which was the Conductor to the infide (
the jar.

" I immediately afked Mr. CONSTA
BLE, if he faw the ball of fire? he fai
Certainly. I faid, I will turn on. F
anfwered, By all means. I kept turnii
the wheel, and the ball of fire continu
turning upon its axis, and afcending up tl
glafs tube till it got quite upon the top
the prime Conductor. There it turn
upon its axis fome little time, and th
gradually defcended, turning upon its a:
as it had done in its afcent, and fo contin
ed till it was fo much below the top of t
coating that we could no longer fee it. F
foon after this, a very great flafh was fec
a large explofion was heard, and ftrong fm
of fulphur was perceived all over the roo
a rou

round aperture was cut through the side
f the jar, as fine as if it had been cut with
. diamond, rather more than three quarters
f an inch in diameter, and between two
nd three inches below the top of the
oating, and the coating was torn off all
ound the aperture, about three or four
nches in diameter. The jar was a pretty
trong one, of crown glafs.

" I then took another jar, fo like the
irft, that when both were whole I could
ot eafily perceive any difference between
hem. I then attempted to charge this
ar, in the fame manner as the other, and
re both obferved it very accurately. No
all of fire was feen, but prefently the jar
ifcharged itfelf with a great flafh and ex-
ofion, and at about the fame part as of
e firft jar; but inftead of the aperture
hich was made in the firft jar, there was
circle about three quarters of an inch
ameter, as white as chalk, and the coat-
g torn off round about it as before.
pon touching the white part, it dropped
t, and appeared to be glafs in a fine
wder.

Q 2   " We

" We broke feveral other different-fized
jars that day, (which made Mr. Consta
ble fay we were in great luck) but with
out any thing elfe remarkable.

" The firft experiment was made fooi
in the afternoon of a clear day, and the ma
chine ftood directly between us and a win
dow, which was not above a yard from it
I don't hear that this ball of fire has beei
produced by art by any one elfe, to thi
day, although it is often produced by na
ture.

" I had the pleafure of feeing Mr. Con
stable this day, and of reading the ac
count of thefe experiments to him, and, t
the beft of his memory, he thought th
whole was ftrictly true.

" Mr. Constable thinks it would no
be difficult to repeat the experiment, and t'
produce the ball of fire at any time, provid
ed the jar is large, and not coated too nea'
the top, and that the wire communicatin
from the prime Conductor to the infide (
the jar is made to pafs through a fma
                                    gla

;lafs tube, (which is certainly of great advantage in making experiments of this kind)
and that the machine acts very ftrong. If
not, it will be in vain to attempt it."

AFTER this letter, Dr. PRIESTLEY,
out of whofe works it has been tranfcribed,
fubjoins the following remarks * :—" The
act mentioned in the preceding letter is
of a very remarkable nature, and, being
perfectly well afcertained, it is of impor-
ance that it be generally known, and kept
in view. For though no perfon, that has
hitherto been made acquainted with it, has
been able to repeat the experiment, others
may be more fortunate. Dr. FRANKLIN,
and, if I miftake not, Mr. CANTON alfo,
and myfelf, were prefent when Mr. HEN-
LY endeavoured to produce this appearance; but, though every expedient that any
of us could fuggeft was made ufe of, we had
no fuccefs, and I have feveral times attempted it in vain fince. I fhall not, however,
defift from my attempts."

* Experiments on Air, &c. vol. v.

Q 3
" Mr.

" Mr. ARDEN's own evidence is abun-
dantly fufficient to authenticate the fact, and
I have fince had from Mr. CONSTABLE
himfelf the fame account of it. Could we
repeat this experiment, there would not, I
think, be any natural phenomenon, in
which the electric fluid is concerned, that
we could not imitate at pleafure. This
circumftance alone makes it a very interest-
ing object of inveftigation."

Nº IV.

OBSERVATIONS CONCERNING SOME PRO-
PERTIES OF THE ELECTRIC FLUID.

AMONGST the various experiments
performed by Mr. WILSON in the
Pantheon, as mentioned in the preceding
pages, he made fome with a long wire,
which was fufpended by filk lines, and
went in various directions through that
great building. This wire was 4800 feet
long, and $\frac{1}{31}$ of an inch in diameter. It
confifted of many pieces, which were connected by twifting the feveral ends together.
—Mr. WILSON electrified this extended
wire by means of an electrical machine fituated next to one of its extremities; and
when the wire was fully charged, which
happened after a few turns of the wheel, a
perfon difcharged it, by prefenting his
hand to it, which was attended with a fhort
fpark, but with a very difagreeable ftrong
fhock. Then Mr. WILSON took an equal
length of wire, of the fame diameter with

Q 4   the

the above-mentioned extended one, and, putting it all coiled together upon a glafs ftand, charged it with electric fluid, by means of the fame machine, and when this wire was fully charged, which took place after about half a turn of the wheel, a fpark was taken from it, but which occa-fioned a fenfation by no means comparable with that given by the extended wire, fince this, though longer than the fpark of the extended wire, was hardly to be felt, where-as that was remarkably ftrong,

The explanation of thofe effects might eafily occur to a perfon little verfed in Electricity. The extended wire acquired a much greater quantity of Electricity than that which was coiled up, for the fame reafon for which an extended piece of tin-foil, twenty fquare feet wide, and which weighs one pound, can acquire incompar-ably more Electricity than a folid ball of tin, which weighs one pound alfo; name-ly, becaufe the extended wire expofes a far greater furface to the free air, than the wire that is coiled up; and the fpark given by the coiled wire is longer than that given

by

by the other, in fo much as the electric
fluid, whatever its quantity may be, can be
much more condenfed on the coiled than
on the extended wire; it being well known,
that the length of the fpark given by electrified Conductors is proportional to the
condenfation, and not to the quantity of
electric fluid.

To obferve, that the extended wire required many turns of the wheel before it
became charged, whereas the coiled wire
required lefs than one turn; the obfervations made and mentioned by various writers on Electricity, relating to the capacity
of Conductors under various circumftances;
Dr. FRANKLIN's experiment of the *can*
and *chain*, and innumerable other experiments of the like nature, might have convinced Mr. WILSON of the truth of the
above-mentioned natural and clear explanation; but, upon thofe phenomena Mr.
WILSON eftablifhes a principle which does
not feem fatisfactory. He imagined, that
in fo much the extended wire gave a much
ftronger fhock than that which was coiled
up, as the electric fluid percurring the

.: whole

whole length of the extended wire acquired a great degree of velocity; hence the impetus with which it ftruck the hand or other body prefented to the charged wire, occafioned that difagreeable fenfation, which fenfation was not accompanied with the fpark given by the coiled wire, becaufe the electric fluid could not acquire any velocity, as it did not go through a great extent of conducting fubftance.

This theory feems to be not true, for the following reafons. Firft, It is well known that metal, and in general any fort of conducting fubftance, does in fome meafure refift the free paffage of the electric fluid through its fubftance; hence the electric fluid, in going through a long Conductor, muft be retarded rather than accelerated. Secondly, The electric fluid being elaftic, whenever a way is open for it to efcape out of a Conductor, upon which it has been confined, it muft exert its greateft force at firft; that is, the quantity of it, which comes out at firft, has the greateft velocity, becaufe it is impelled by the greateft expanfive force; but, according as more and more of the electric fluid comes out of the Conductor, that impelling force or elasticity leffens ; hence the laft particles of electric fluid come out of the Conductor with lefs velocity ; thus in a condenfing fountain, out of which the ftream of water iffues, on account of the elafticity of the air condenf-ed in the cavity of the inftrument, the ftream at firft comes out with greater vio-lence, and mounts higher up; but after-wards lofes its velocity by degrees, becaufe the condenfed air becomes gradually lefs and lefs condenfed, and confequently exerts a force continually fmaller and fmaller againft the water. Laftly, It may be ob-ferved in general, as it has been proved by many experiments, that if two unequal Conductors, one of which A is more ex-tended than the other B, difcharge equal quantities of electric fluid under the form of fparks from like terminations, the fpark given by A is lefs ftrong than the fpark given by B, and that for the above-men-tioned reafons.

Notwithftanding the obvioufnefs of thofe remarks, Mr. WILSON, fince he firft made

the

the experiments at the Pantheon, feems to
have become continually more and more
attached to his opinion, fo that fome time
ago, he publifhed a fmall tract, entitled
A fhort View of Electricity, in which he
endeavours to eftablifh his theory of the
acceleration of the electric fluid, &c. upon
mathematical principles. This ftep determined me to take notice of it in this Appendix, in order to prevent any wrong no-
tions in the heads of novices ; for perfons
who are not converfant with mathematics,
will take any propofition for true, when
they are affured that it is mathematically
fo. Mr. WILSON's theorem conceals a
fallacy, and fhews that a perfon very well
fkilled in a fcience, as Mr. WILSON is in
Electricity, can eafily fall into a miftake,
when he is too much attached to a favou-
rite opinion.—Mr. WILSON's words are
the following :

" Upon Acceleration.

" The experiments at the *Pantheon,*
" which were intended to fhew the acce-
" leration of the fluid, having been object-
" ed to by many, who have not fufficiently
" attended

" attended to the known properties of the
" elaftic fluid, it has been thought proper
" here to eftablifh this very material point
" upon mathematical principles, with a
" view to put an end to all farther dif-
" putes on the fubject.

" But before this is done, it may be
" neceffary to mention·a material fact that
" was omitted in the account of thofe ex-
" periments ; which is this,

" The fhock received at the *middle* of
" the long wire, was *confiderably* lefs than
" that which was received at either end.

" T H E O R E M.

A   B

1   2   3   4   5, &c.   N

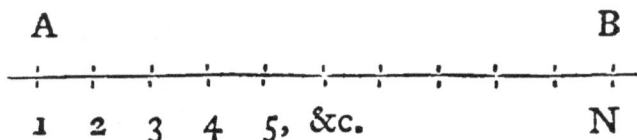

" Let A B reprefent a cylinder of a given
" diameter, and fuppofe this cylinder
" charged with the electric fluid. I fay,
" if all the particles of this fluid are moved
" at the fame inftant towards A, the effect
" produced by the fhock of this fluid at
                   " A will

" A will be nearly proportional to the
" fquare of A B; for the total effect at
" A is equal to the fum of the effect of each
" particle contained in the cylinder A B:
" and the effect of each particle being pro-
" portional to its velocity, the total effect
" at A will be proportional to the fum of
" all the velocities.. But fince the fluid is
" fuppofed. nearly perfectly elaftic, all the
" particles will arrive at A nearly at the
" fame inftant; then the velocity of each
" particle will be proportional to the dif-
" tance from the place it fets out, and the
" total effect at A will be proportional to
" the fum of all thofe diftances.

" But all thofe diftances are exprefled
" by the following numbers, 1, 2, 3, 4, 5,
" &c. - - - - - - - N (N exprefling the
" length A B) in an arithmetical progref-
" fion. Then the fum of all the diftances
" will be exprefled by the fum of the
" arithmetical progreffion 1, 2, 3, 4, 5,
" &c. - - - - - - - N, and the effect at A
" will be proportional to this fum, that is
" to fay, to N^2 or AB^2. Q. E. D."

Mr.

Mr. Wilson's abstract proposition is certainly evident, *viz.* " That if all the particles of the electric fluid are moved at the same instant towards A, (by which, I suppose, he means if they arrive at A in the same instant of time) the effect produced by the shock of this fluid at A will be nearly proportional to the square of A B." But this proposition cannot be possibly applied to the matter of fact, or actual experiment, since the condition upon which the proposition depends cannot be verified ; *viz.* we do not know whether or no the particles of the electric fluid that are in the long wire, do all arrive at one end of it precisely at the same time. Besides, we may safely say, that from the considerations mentioned in the preceding pages, and from the analogy of other natural phenomena, it appears that they do not, and cannot arrive to the extremity of the wire precisely at the same time. Between the particles of electric fluid which arrive first at one extremity of the wire, and those which arrive last at the same extremity, there is an interval, a difference of time. Mr. Wilson says, " all the particles will arrive at
A, *nearly*

A, *nearly* at the fame inftant." Now, even
that *nearly* implies a length of time, which,
however imperceptible it may be to our
fenfes, is yet fufficient for the natural operation. For inftance, fuppofe that the extended
and electrified wire is a mile long, and alfo
fuppofe that a quantity of electric fluid em-
ploys one hundredth part of a fecond of time to
go through that length of wire, and that without acceleration, or even with a little retarda-
tion, which is by no means an exaggerated
fuppofition. Now, if a way be opened to the
electric fluid in the wire, *viz.* the hand, &c.
be prefented to one extremity of it, the
neareft particles of the electric fluid will
come out firft, and the remoteft laft; yet
that difference, agreeably to the fuppofition,
does not exceed one hundredth part of a fecond, which is abfolutely unperceivable by
our fenfes; fo that the appearance is exactly
the fame as if all the particles of the electric fluid came out of the extremity of the
wire precifely at the fame inftant of time.
I think, therefore, that what Mr. WILSON
imagines to have eftablifhed upon mathematical principles, is far from being mathematically demonftrable. Without dwelling
any

any longer upon Mr. WILSON's propofition, and the corollaries which he deduces from it, I fhall only take notice of a remarkable experiment he made at the Pantheon, with the already-mentioned apparatus ; which, however, he attributes to the accumulated velocity, which the electric fluid acquired in percurring the fubftance of a very long Conductor.

Mr. WILSON fired gunpowder by means of the ftream of electric fluid, but without fpark or fhock. The method is as follows, in his own words : " Upon a ftaff of baked " wood a ftem of brafs was fixed, which " terminated in an iron point at the top. " This point was put into the end of a " fmall tube of Indian paper, made fome- " what in the form of a cartridge, about " two-tenths of an inch in diameter. " When this cartridge was filled with " common gunpowder (unbruifed) the " wire of communication with the well " was then faftened to the bottom of the " brafs ftem. Being fo circumftanced, and " whilft the charge in the great cylinder " and wire was continually kept up by

VOL. II.   R   " the

" the motion of the wheel, the top of the
" cartridge was brought fo near to the
" drums as frequently to touch the metal.
" In this fituation, a fmall, faint, lumi-
" nous ftream was obferved between the
" top of the cartridge and the metal
" drum.

" Sometimes this ftream would fet fire
" to the gunpowder at the inftant of the
" application; at others, it would require
" half a minute, or more, before it took
" effect. But this difference in time
" might probably arife from fome differ-
" ence in the circumftances; for any the
" leaft moifture, &c."

It feems that in this experiment the gunpowder was fired by the heat occafioned
by the denfe ftream iffuing out of the vaft
Conductor, which difcharged a prodigious
quantity of electric fluid in that concentrated manner. This experiment was afterwards imitated by Mr. NAIRNE, with
the Leyden phial. When a large electric
jar was charged, Mr. NAIRNE prefented
the cartridge of gunpowder, fimilar to that

ufed

ufed by Mr. WILSON, to the knob of the jar, and the powder was fired without any explofion of the jar. In this experiment, the wire upon which the cartridge was fixed, did not communicate with the outfide of the jar by a good conducting communication, but both the outfide of the jar, and the faid wire, did communicate with the ground; fo that it may be faid, that the circuit between the two fides of the jar was imperfect; hence the difcharge was not made by a full explofion.

EXTRACTS FROM MR. VOLTA'S PAPER,

Inferted in the 72d vol. of the Phil. Tranf.

Concerning the Capacity of Conductors, a new Method of difcovering very fmall Degrees of Electricity, and the Electricity of the Atmofphere.

MR. VOLTA'S paper on the method of difcovering the fmalleft degrees of Electricity, either natural or artificial, containing feveral particulars not only new, but alfo very interefting to electricians, it was deemed neceffary to infert in this book at leaft a fummary account of them; and had the limits of the work permitted it, I fhould have tranfcribed the whole paper; however, the extracts are fuch as, I imagine, will convey a diftinct idea of the difcoveries of that excellent electrician, and will leave out not more than what the ingenuity of the reader may eafily fupply. I have generally ufed the
word

words of Mr. VOLTA himfelf; but I have changed in great meafure the order of the paragraphs, for the fake of rendering the fubject more intelligible in a contracted ftate.

Conducting bodies, of the fame fhape, will contain a greater or lefs quantity of Electricity, according as their furfaces are lefs or more influenced by *homologous atmofpheres*; and the capacity of a Conductor, which has neither its form nor furface altered, is increafed when, inftead of remaining quite infulated, the Conductor is prefented to another Conductor not infulated, and this increafe is more confpicuous, according as the furfaces of thofe Conductors are larger and come nearer to each other.—When an infulated Conductor is oppofed or prefented to another Conductor whatever, Mr. VOL-TA calls it a *conjugate Conductor*.

In order to fhew by experiment the above-mentioned property or increafe of capacity in a Conductor, take the metal plate of an electrophorus, and holding it by its infulating handle in the air, electrify it fo high, as that the index of an electrometer an-

R 3   nexed

nexed to it might be elevated to 60°; then lowering this metal plate by degrees towards a table or other conducting plain surface, you will observe that the index of the electrometer will fall gradually from 60° to 50°, 40°, 30°, &c. Notwithstanding this appearance, the quantity of Electricity in the plate remains the same, except the said plate be brought so near the table as to occasion a transmission of the Electricity from the former to the latter; at least the quantity of Electricity will remain as much the same as the dampness of the air, &c. will permit. The decrease, therefore, of intensity is owing to the increased capacity of the plate, which is now *conjugate, viz.* opposed to another conducting surface. In proof of which, remove gradually the metal plate from the table, and it will be found that the electrometer rises again to its former station, namely to 60°, excepting the loss of that quantity of Electricity, which during the experiment must have been imparted to the air.

The reason of this phenomenon is easily derived from the action of electric atmospheres.

fpheres. The atmofphere of the metal plate, which, for the prefent, we fhall fuppofe to be electrified pofitively, acts upon the table or other Conductor to which it is prefented; fo that the electric fluid of the table, agreeably to the known laws, retiring to the remoter parts of it, becomes more rare in thofe parts which are expofed to the metal plate, and this rarefaction becomes greater, the nearer the electrified metal plate is brought to the table.   If the metal plate is electrified, then the contrary effects muft take place. In fhort, the parts immerfed into the fphere of action of the electrified metal plate, contract a contrary Electricity, which *accidental* Electricity, making in fome manner a compenfation for the *real* Electricity of the metal plate, diminifhes its intenfity, as is fhewn by the depreffion of the electrometer.

The two following experiments will throw more light upon the reciprocal action of the electric atmofpheres.   Firft, fuppofe two flat Conductors, electrified both pofitively or both negatively, to be prefented towards, and to be gradually brought near,

each other; it will appear by two annexed
electrometers, that the nearer those two
Conductors come to each other, the more
their intensities will increase; which shews,
that either of the two *conjugate* Conductors
has a much less capacity now, than when it
was singly insulated, and out of the influence
of the other, This experiment explains
the reason why an electrified Conductor
will shew a greater degree of intensity when
it comes to be contracted into a smaller bulk;
and also why a long extended Conductor
will shew a less degree of intensity than a
more compact one, supposing that the quantity of surface and of Electricity is the same
in both; because the homologous atmospheres of their parts interfere less with
each other in the former than in the latter
case.

Secondly, let the preceding experiment be
repeated, with this variation only, *viz.* that
one of the flat Conductors be electrified positively, and the other negatively: the effects then will be juft the reverse of the
preceding; *viz.* the intensities of their
Electricities will be diminished, because
their

their capacities are increafed, the nearer the Conductors come to each other.

Let us now apply the explanation of this laft experiment to that of bringing an electrified metal plate towards an uninfulated conducting plane; for as this plane acquires a contrary Electricity by the vicinity of the electrified plate, it follows that the intenfity of the Electricity of the metal plate muft be diminifhed, and in the fame proportion its capacity is increafed; confequently the metal plate in that cafe may receive a greater quantity of Electricity.

This property may be rendered ftill more evident, by infulating the conducting plane whilft the electrified plate is very near it, and afterwards feparating them; for then both the metal plate and the conducting plane (which may be called the *inferior* plane) will be found electrified, but poffeffed of contrary Electricities, as may be afcertained by electrometers.

If the inferior plane be infulated firft, and then the electrified plate be brought over it,
then

then the latter will cause an endeavour in the former to acquire a contrary Electricity, which however the insulation prevents from taking place; hence the intensity of the Electricity of the plate is not diminished, at least the electrometer will shew a very little and almost imperceptible depression, which is owing to the imperfection of the insulation of the inferior plane, and to the small rarefaction and condensation of the electric fluid, which may take place in different parts of the said inferior plane. But if in this situation the inferior plane be touched, so as to cut off the insulation for a moment, then it will immediately acquire the contrary Electricity, and the intensity in the metal plate will be diminished.

If the inferior plane, instead of being insulated, were itself a non-conducting substance, then the same phenomena would happen, *viz.* the intensity of the electrified metal plate laid upon it would not be diminished. This, however, is not always the case; for if the said inferior non-conducting plane be very thin, and be laid upon a Conductor, then the intensity of the electrified

metal

metal plate will be diminifhed, and its capacity will be increafed by being laid upon
the thin infulating ftratum ; becaufe in that
cafe the conducting fubftance, which ftands
under the non-conducting ftratum, acquiring an Electricity contrary to that of the
metal plate, will diminifh its intenfity, &c.
and the infulating ftratum will only diminifh the mutual action of the two atmofpheres, more or lefs, according as it keeps
them more or lefs afunder.

The intenfity or electric action of the
metal plate, which diminifhes gradually as
it is brought nearer and nearer to a conducting plane not infulated, becomes almoft
nothing when the plate is nearly in contact
with the plane, the compenfation or accidental balance being then almoft perfect ;
hence if the inferior plane only oppofes a
fmall refiftance to the paffage of the Electricity (whether fuch refiftance be occafioned
by a thin electric ftratum, or by the plane's
imperfect conducting nature, as is the cafe
with dry wood, marble, &c.) that refiftance, and the interval, however fmall, that
is between the two planes, cannot be overcome

come by the weak intenfity of the Electricity of the metal plate, which on that account will not dart any fpark to the inferior
plane (except its Electricity were very
powerful, or its edges not well rounded)
and will rather retain its Electricity; fo
that, being removed from the inferior plane,
its electrometer will nearly recover its former height. Befides, the electrified plate may
even come to touch the imperfectly conducting plane, and may remain in that fituation for fome time : in which cafe the intenfity being reduced almoft to nothing, the
Electricity will pafs to the inferior plane
exceedingly flowly.

But the cafe will not be the fame, if, in
performing this experiment, the electrified
metal plate be made to touch the inferior
plane *edgewife*; for then its intenfity being
greater than when laid flat, as it appears by
the electrometer, the Electricity eafily over-
comes the fmall refiftance, and paffes to
the inferior plane, even acrofs a thin electric ftratum ; becaufe the Electricity of
one plane is balanced by that of the other,
only in proportion to the quantity of furface

face which they oppofe to each other within a given diftance; whereby, when the
metal plate touches the other plane in flat
and ample contact, its Electricity is not diffipated.

This explanation, properly applied, renders evident the action of *points* in general.
Juftly fpeaking, a pointed Conductor, not
infulated, when prefented to an electrified
body, has not in itfelf any particular virtue
of attracting Electricity.   It acts only like
a Conductor not infulated, which does not
oppofe any refiftance to the paffage of the
electric fluid.   If the fame Conductor, inftead of being pointed, were to prefent a
globular or flat furface to the electrified
body, neither in that cafe it would oppofe a
greater refiftance to the paffage of the Electricity.   But the reafon why the Electricity
will not pafs nearly fo eafily from the electrified body to the globular or flat conductor, as to the pointed one, is becaufe in the
former cafe the intenfity of the Electricity in
the electrified body is weakened by the oppofed flat furface, which, acquiring the
contrary Electricity, compenfates the diminifhed

minifhed intenfity incomparably more than a point, is able to do. It appears, therefore, that it is not the particular property of a point or of a flat furface, but the different ftate of the electrified body, that makes it part with its Electricity eafier, and from a greater diftance, when a pointed conducting fubftance, than when a flat globular one, is prefented to it.

What looks like a paradox in the cafe of an electrified plate ftanding upon an imperfectly conducting plane, is, that the metal plate, whilft ftanding upon the other plane, will not lofe all its Electricity, even if it be touched with a finger or with a piece of metal; fo that it generally remains fo far electrified, as when it is afterwards feparated from that plane, it will often afford a fmall fpark, or at leaft it will affect an electrometer. Indeed this phenomenon could not be explained upon the fuppofition that the finger or the metals were perfect Conductors : but fince we do not know of any perfect Conductor, the metals or the finger oppofe a refiftance fufficient to retard the immediate diffipation of the Electricity of

of the plate, which is in that cafe actuated
by a very fmall degree of intenfity or endeavour of expanding; fo that fuppofe, for
inftance, that the piece of metal or the finger, by touching the plate, took off fo much
of its Electricity as to reduce the intenfity
of the remainder to the fiftieth part of a
degree, this remaining Electricity would
then be almoft nothing; but when the
plate, by being feparated from the inferior
plane, has its capacity fo far diminifhed as
to render the intenfity of its Electricity one
hundred times greater, then the intenfity of
that remaining Electricity would become of
two degrees, viz. fufficient to afford a fpark
or to affect an electrometer.

Hitherto we have confidered in what
manner the action of electric atmofpheres
muft modify the Electricity of the metal
plate in various fituations. We muft now
confider the effects which take place when
the Electricity is communicated to the metal plate whilft ftanding upon the imperfectly
conducting plane; however the explanation
of this eafily follows from what has been
faid above. Suppofe, for inftance, that a
                                    Leyden

Leyden phial or a Conductor were so weakly electrified, that the intensity of its Electricity were only of half a degree, or even less; if the metal plate, when standing upon the proper plane, were touched with that phial or Conductor, it is evident that either of them would impart to it a quantity of its Electricity, proportional to the plate's capacity, *viz.* so much of it as would make the intensity of the Electricity of the plate equal to that of the Electricity in the Conductor or phial, supposed of half a degree; but the plate's capacity, now that it lies upon the proper plane, is above one hundred times greater than if it stood insulated in the air; or, which is the same thing, it requires one hundred times more Electricity in order to shew the same intensity; therefore, in this case, it must acquire upwards of a hundred times more Electricity from the phial or Conductor. It naturally follows, that when the metal plate is afterwards removed from the proper plane, its capacity being lessened so as to remain equal to the hundredth part of what it was before, the intensity of its Electricity must become of 50°; since, agreeably to the supposition, the

intensity

intenfity of the Electricity in the phial or Conductor was of half a degree *.

A Conductor that is electrified whilft ftanding in full and ample contact with another proper Conductor, or femi-infulating body, as above mentioned, and is afterwards feparated from it, fhews the fame phenomena that are exhibited by a Conductor, which, after being electrified, is contracted into a fmaller bulk, or contrarywife, like Dr. FRANKLIN's experiment of the can and chain, &c. †.

If a fmall quantity of Electricity communicated to the metal plate, whilft ftand-

---

* In a paper on the capacity of fimple Conductors, Mr. VOLTA, confidering the great capacity of a Leyden phial, fhews that the caufe of it is the Electricity communicated to one of its furfaces being balanced by the contrary Electricity of the oppofite furface. He likewife proves, that the capacity of fixteen fquare inches of coated electric furface is equal to the capacity of a Conductor made of filvered cylindrical fticks, nearly one hundred feet long, the capacity of which is fo great as to afford a fpark capable of occafioning a fhock confiderably ftrong.

† See page 331, vol. I. of this work.

ing.

ing on the proper plane, will afterwards enable it to give a ftrong fpark, it may be afked, what would a great quantity of Electricity do ? The anfwer is, that it would do nothing more; becaufe, when the Electricity communicated to the metal plate is fo ftrong as to overcome the fmall refiftance of the inferior plane, it will then be diffipated.

It will be readily underftood, that if the metal plate fituated upon a proper plane can receive a good fhare of Electricity from a Leyden phial, or from an ample Conductor, however weakly electrified, it cannot receive any confiderable quantity of it from a Conductor of a fmall capacity; for this Conductor cannot give what it has not, except it were continually receiving a ftream, howfoever fmall, of Electricity, as is the cafe with an atmofpherical Conductor, or with the prime Conductor of an electrical machine, which acts very poorly, but continues in action. However in fuch cafe a confiderable time muft elapfe before the metal plate has acquired a fufficient quantity of Electricity.

Thus

Thus much being premifed relating to the capacity of Conductors, we now proceed to defcribe Mr. VOLTA's ingenious method of difcovering, or of rendering fenfible, the fmalleft quantity of Electricity, which is entirely dependant on the foregoing principles.

The method, in fhort, is to communicate the fmall, and otherwife unobfervable quantity of Electricity, to the metal plate of an electrophorus whilft ftanding on an imperfectly infulating plane; for in that fituation the capacity of the metal plate being augmented, it will acquire an incomparably greater quantity of Electricity than if it ftood infulated in the air; and afterwards, when feparated from the plane, its capacity will be contracted; and confequently, its Electricity increafing at the fame time, its intenfenefs will evidently manifeft itfelf either by fparks, or, which is the eafieft and fafeft method, by affecting an electrometer.

The particulars neceffary to be kept in view in this method are the following. The metal plate muft be at leaft fix inches in di-

S 2   ameter,

ameter, with the edge well rounded, and having a varnifhed glafs handle ;—inftead of a glafs fome perfons have ufed three filk ftrings. The inferior plane muft be of a very imperfect conducting nature, as dry marble, very dry and flightly varnifhed wood, a common piece of wood covered with oiled filk, or fuch-like fubftance ; but let the fubftance be what it will, its furface muft be very fmooth, and fuch as to coincide as well as poflible with the furface of the metal plate; on which account, if a marble flab be chofen for the inferior plane, it will be proper to fit the furface of the metal plate to that of the marble, by grinding one againft the other. What I find to be very fit for this purpofe is a paper drum, confifting of a common wooden hoop, fuch as are ufed for cafks, over which a piece of thick writing-paper is pafted, on the back of which I pafte a piece of tin-foil. The upper furface of the paper is varnifhed only once with fhell-lac diffolved in fpirit of wine. This fort of plane has many advantages, *viz.* it is eafily made, and being light is very portable; its furface is perfectly plane, excepting indeed when the

hoop

hoop is not very ſtrong, for then the contraction of the paper has power ſufficient to
warp it ; and laſtly, as the thickneſs of the
paper and of the varniſh may be varied at
pleaſure, and very eaſily, the ſaid plane may
be rendered of any required degree of conducting power.

Now ſuppoſing that an inferior ſemi-con-
ducting plane and a metal plate are properly
conſtructed, lay the former upon a table, and
upon it lay the latter, taking care that in
this operation, or in wiping, &c. the ſurface
of the inferior plane be not excited, for
that would fruſtrate the experiment*. Then,
with the corner of a dry handkerchief, a
piece of flannel, of paper, &c. ſtrike five or
ſix times the metal plate. Afterwards,
laying hold of the glaſs handle, -lift up the
metal plate from over the inferior plane,
and preſenting it to the electrometer you
will find that it is evidently electrified. If

* In caſe the ſurface of the inferior plane has
acquired any Electricity, either from rubbing it inad-
vertently, orotherwiſe, the beſt way of freeing it of that
Electricity, is to paſs it over the flame of a candle two
or three times.

this

this experiment of ſtriking the plate be repeated when the plate is not on the proper plane, you will find either not the leaſt veſtige of Electricity, or a quantity of it incomparably ſmaller than that obtained in the other way .

By this means, Mr. VOLTA has obtained Electricity not only ſufficient to manifeſt its quality, but even enough to afford ſparks, and that from ſubſtances which could be hardly ſuſpected to be electrified. An atmoſpherical Conductor not much elevated above the top of a houſe, if it be made to communicate with the metallic plate ſtanding on the proper plane (which Mr. VOLTA juſtly calls the *electrical condenſing apparatus*) will be found to be electrified at times in which it would not ſhew any ſigns of Electricity by any other means. Several ſubſtances, even ſome that are conſidered as very good Conductors, will be found to afford a ſenſible quantity of Electricity, when the metal plate of the condenſing apparatus is ſtroked with any of them, ſuch as pieces of cloth, or of leather, moſt green vegetables, and even the human hand,

hand. Indeed there is hardly any fubftance, befides fluids or very foft fubftances, which will not afford a fenfible degree of Electricity by this method. But the moft remarkable difcoveries made by Mr. VOLTA, are that the evaporation of water and other fluids produces Electricity. Some effervefcences he alfo found to produce Electricity, though their Electricity may be only the confequence of the evaporation which generally accompanies effervefcences; but before I mention any thing farther relating to thofe difcoveries, I fhall juft defcribe two improvements of mine concerning the abovementioned condenfing apparatus.

Obferving that in ftroking the metal plate in order to obtain Electricity from various fubftances, and efpecially from the hand, the plate was often moved fo as to occafion fome friction upon the inferior plane, which fometimes excited that plane in a fmall degree, and confequently rendered precarious the refult of the experiment, I thought of the following improvement, which entirely prevents any motion being communicated to the metal plate.

Upon

Upon a varnished glass handle a brass tube about six inches long, and $\frac{1}{4}$ of an inch in diameter, was cemented, and from the extremity of this tube a fine and very flexible wire proceeded, which was about fourteen inches long. Now when the metal plate was situated upon the inferior plane, I held the glass handle of the brass tube with my left hand, in such a manner as that the end of the wire might touch the plate, the rest remaining in the air. Sometimes, in order to make a better contact, the end of the above-mentioned wire was put into a hole purposely made in the edge of the plate. In this disposition of the apparatus the substances to be tried are stroked upon the brass tube, and the Electricity produced by them is conveyed to the metal plate by the wire, which being fine and flexible communicates not the least motion to the said plate.

The other improvement consists in rendering sensible degrees of Electricity still smaller than those, which may be discovered by the above-mentioned condensing apparatus. This improvement was suggested by, and is founded upon, the same principles,

Notwith-

Notwithstanding the great sensibility of this apparatus of Mr. VOLTA, yet sometimes the Electricity acquired by the metal plate from some substances, was so small as not to affect an electrometer sufficiently to ascertain its quality, or even its existence; hence I naturally thought that, for the same reason for which the metal plate of the condensing apparatus manifested such small degrees of Electricity as could not be otherwise observed, another smaller plate, or small condensing apparatus, might be employed to collect and render sensible the weak Electricity of the large metal plate. Accordingly I constructed a small plate of about the size of a shilling, having a glass handle covered with sealing-wax; and when the large metal plate seemed to be so weakly electrified as not to affect an electrometer sensibly, I placed the small plate upon the inferior plane, and touched it with the edge of the large plate: then, after removing the large plate, I took up the small one from the plane, holding it by the extremity of the glass handle, and presented it to the electrometer, which was generally so much affected by it as to diverge to its utmost limits.

In

In this manner I have often obtained Electricity more than sufficient to ascertain its quality, from a single stroke of the corner of a handkerchief: *viz*. the large plate being placed upon the proper plane was stroked once; then being removed and presented to an electrometer, it appeared not electrified, but by touching the small plate with the edge of it, that small plate acquired thereby Electricity sufficient to make an electrometer diverge.

When this secondary condensing apparatus is used, care must be had to hold the large plate almost vertically whilst the small plate is touched with it. There is no need of having another inferior plane for the small plate, the large one being sufficient for both; for immediately after taking up the large plate weakly electrified, with one hand, you lay down the small plate, &c.

The little quantity of Electricity that can be discovered by this means is really surprising; and there is hardly any substance, excepting the metals, or those which cannot be
subjected

ſubjeſted to trial, as water and other fluids, which will not produce ſome Electricity when rubbed or ſtroked againſt the large plate of the condenſing apparatus, and that Electricity is afterwards condenſed by being communicated to the ſmall plate. In conſequence of thoſe experiments it appears, that throughout the works of nature there is a continual motion of the electric fluid from one ſubſtance to another ; ſince there is hardly any friction, and, as it will appear by what follows; any evaporation or condenſation, but it produces Electricity. The extenſive influence of this fluid, or of that power which we call *electric*, and the immenſe dependence of the powers of nature on each other, indicate that Electricity muſt be employed for ſome great operation ; but human induſtry has not yet removed the veil from this great myſtery.

The principal diſcovery, which Mr. VOLTA made by means of the condenſing apparatus, is the Electricity produced by the evaporation of water, which explains in a great meaſure, if not entirely, the origin of
the

the atmofpherical Electricity*. He found that water quickly evaporated, that the fimple combuftion of coals, and that the effervefcence of iron filings, in diluted vitriolic acid, when performed in infulated veffels, left the veffels negatively electrified. This negative Electricity he generally collected and rendered manifeft by means of the above-defcribed condenfing apparatus ; but fometimes it is fo ftrong as not to require any thing more than a fenfible electrometer connected with the infulated veffel by means of Conductors, as a wire or the like.

The experiment of producing Electricity by the evaporation of water may be eafily performed, thus:—Upon an infulating ftand, as a wine glafs or other electric fubftance, place an earthen veffel, as a crucible, a bafin, or fuch like thing, and put into it three or four lighted coals. Let a wire be put with one end amongft the coals, and with the other let it touch a very fenfible electrometer. (One of my improved atmo-

* This difcovery was made on the 13th of April, 1782.

fpherical electrometers, *viz.* that in the bottle, anfwers very well.) Then pour a fpoonful of water at once upon the coals, which will occafion a great hiffing, and a quick evaporation; and at the fame time you will fee the electrometer diverge with negative Electricity.

Mr. VOLTA concludes the account of thofe difcoveries with the following inge-nious remarks. " The experiments," *fays he,* " hitherto made, though not numerous, " yet concur to fhew, that the vapours of " water, and in general the parts of all bo- " dies, that are feparated by volatilization, " carry away an additional quantity of elec- " tric fluid as well as of elementary heat; " and confequently that thofe bodies, from " the contact of which the volatile parti- " cles have been feparated, remain both " cooled and electrified negatively: from " which it may be deduced, that whenever " bodies are refolved into volatile elaftic " fluid, their capacity for holding electric " fluid is augmented, as well as their ca- " pacity for holding common fire, or the " calorific fluid. This is a ftriking analogy " by

" by which the fcience of Electricity
" throws fome light upon the theory of
" heat, and alternately derives light from
" it.

" By following this analogy it feems,
" that as the vapours on their condenfing
" lofe part of their latent heat, on account
" of their capacity being diminifhed, fo
" they part with fome electric fluid. Hence
" originates the pofitive Electricity, which
" is always more or lefs predominant in the
" atmofphere, when the fky is clear, viz.
" at that height where the vapours begin
" to be condenfed. Accordingly, the at-
" mofpherical Electricity is ftronger in fogs,
" in which cafe the vapours are more con-
" denfed, fo as to be almoft reduced into
" drops, and is ftill ftronger when thick
" fogs become clouds.

" Hitherto we have accounted for the
" pofitive atmofpherical Electricity; but it
" is eafy to account for clouds negatively
" electrified; for when a cloud, pofitively
" electrified, has been once formed, its
" fphere of action is extended a great way
" round,

" round, fo that if another cloud comes
" within that fphere, its electric fluid,
" agreeably to the well-known laws of
" electric atmofpheres, muft retire to the
" parts of it which are the remoteft from
" the firft cloud ; and from thence the
" electric fluid may be communicated to
" other clouds, or vapours, or terreftrial
" prominencies. Thus a cloud may be
" electrified negatively, which cloud, after
" the fame manner, may occafion a pofitive
" Electricity in another cloud, &c. This
" explains not only the negative Electricity,
" which is often obtained from the atmo-
" fphere in cloudy weather, and the fre-
" quent changes from pofitive to negative
" Electricity, and contrarywife in ftormy
" weather, but alfo the waving motion
" often obferved in the clouds, and the
" hanging down of them, fo as nearly to
" touch the earth.

" After the fore-mentioned difcoveries,
" we need no longer wonder at the appear-
" ance of lightnings in the eruptions of vol-
" canos, as was particularly obferved in the
" late

" late dreadful eruption of Mount Vefuvius.
" The few experiments I have made, fhew
" that the quantity of fmoke, but much
" more the rapidity with which it is pro-
" duced, tends to increafe the Electricity,
" which arifes from combuftion, &c. How
" great muft then be the quantity of Elec-
" tricity that is produced in fuch erup-
" tions ! "

N° VI.

N° VI.

AN ACCOUNT OF A VERY POWERFUL
ELECTRICAL MACHINE, LATELY CON-
STRUCTED FOR THE MUSEUM OF TEY-
LER, AT HARLEM; AND OF VARIOUS
EXPERIMENTS MADE WITH IT.

THE defcription of this machine, and
of feveral experiments made with it,
was written by Mr. VAN MARUM, who
has himfelf been the principal operator in
thofe experiments. The various new and
interefting particulars which are mentioned
in this defcription, render it deferving of being read at large by every lover of philofo-
phy; and the prefent fhort extract is intended only to give an idea of it to thofe who
have not the opportunity of perufing the
original, which was lately publifhed at
Harlem.

This machine confifts of two circular
glafs plates, each 65 inches in diameter,
which being fixed parallel to each other, and

7¼ inches afunder, on a common axis, are turned by a winch without any accelerated motion, and are rubbed by eight rubbers, all being placed in a proper frame. Each plate is rubbed on both furfaces; two cufhions or rubbers being on one fide, and two on the oppofite fide of each plate. The prime Conductor is divided into branches, which enter between the plates, and, by means of points, collect the electric fluid only from their inner furfaces.

In general two men are employed to work this machine, but when it is required to keep it in action for a long time, then four men are put to it.

The power of this machine appears to be greater than that of any other made before, and indeed its effects are furprifing. I fhall only enumerate them as briefly as poffible, without expatiating on the points of comparifon between them and others of a fimilar nature, performed with other machines.

A very fharp fteel point, being prefented to the prime Conductor, drew a luminous
ftream

ftream of electric fluid, about half an inch long.

When a fharp fteel point was fixed to the Conductor, fo as to project three inches from it, on working the machine the point threw out ftreams of light, which were fix inches long when a ball of three inches in diameter was prefented to it, and only two inches long when another point was prefented in-ftead of the ball.

The fenfation, commonly called the fpi-der's web, on the face of the by-ftanders, when this machine is in action, is often felt at the diftance of eight feet from the prime Conductor.

A thread fix feet long, fufpended perpen-dicularly, was fenfibly attracted by the prime Conductor at the diftance of thirty-eight feet. This attraction of the thread by the prime Conductor was fuch as juft to make the thread deviate from the perpendicular direction. A pointed wire, prefented to the Conductor, appeared luminous, even when the diftance between the former and the latter was twenty-eight feet.

Another

Another Conductor being prefented to the prime Conductor, in order to receive the fparks from it, and a perfect metallic communication being made between the former and the earth, by means of a long brafs wire $\frac{1}{4}$ of an inch in diameter; it was found, that whilft a ftream of electric fluid went from the prime Conductor to the other, which may be called the *receiving Conductor*, the brafs wire gave fmall fparks to conducting bodies that were placed near it. The quantity of electric fluid, accumulated by this machine and difcharged by the prime Conductor, muft have been really furprifingly great; fince a wire of $\frac{1}{4}$ of an inch in diameter was not capable of tranfmitting it to the earth very eafily. The fparks between the two Conductors were generally 21, but fometimes 24 inches long. The ftream of fire was crooked, and darted many lateral brufhes of a very large fize.

Gunpowder, after the manner formerly ufed by Mr. WILSON, as it is mentioned in the preceding pages, and fome other combuftibles, could be fired at the prime Conductor of this machine.

A fingle

A single spark from the Conductor melted a considerable length of gold-leaf.

A Leyden phial, containing about one square foot of coated surface, was fully charged by about half a turn of the winch, so as to discharge itself; and by repeated trials it was found, that in one minute's time this phial discharged itself 76, 78, and often 80 times.

By many repeated experiments, accurately made by Mr. VAN MARUM, and other ingenious persons, with this machine, it was found, that electrization, whether positive or negative, did neither sensibly augment nor diminish the natural number of pulsations in a man *.

They

* In my Essay on Medical Electricity it is mentioned, that from the experience of many it appeared, that Electrization increases the number of pulsations about one sixth; but having made several experiments upon myself, I added the following observation in the second edition of the said Essay, which was published in the year 1781, and consequently long before Mr. VAN MARUM's experiments. " I do not remember that " my pulse was ever evidently accelerated by electri-

T 3   " zation,

They wifhed to try the effect of taking the electric fparks in different forts of elaftic fluids, and for this purpofe, they ufed a cylindrical glafs receiver, five inches long, and an inch and a quarter in diameter, into which different forts of elaftic fluids were fucceffively inferted, and were confined by quickfilver or water. To a hole made in the bottom of the inverted glafs receiver an iron wire was faftened, the external part of which communicated with a Conductor, which, being prefented to the prime Conductor of the machine, received the fparks from it. In this difpofition of the apparatus it evidently appears, that the fparks paffed through the elaftic fluid contained in the receiver, by going from the inner extre-

"zation, and yet I have repeated the experiment va-
"rious times, and with great diverfity of circumftan-
"ces." P. 15.—Upon the whole therefore it feems to be afcertained, that electrization does not accelerate nor retard the ordinary number of pulfations, and that the augmentation generally obferved before muft have been owing to fear or apprehenfion. But I am informed by Mr. PARTINGTON, who has long practifed medical Electricity, that electrization, if not in a found, at leaft in an unfound ftate of the body, augments the number of pulfations confiderably.

mity of the wire to the quickfilver or water in which the receiver was inverted. With this apparatus it was found, that dephlogifticated air, obtained from mercurial red precipitate, loft $\frac{1}{20}$ of its bulk, but its quality was not fenfibly altered, as it appeared from examining it with the eudiometer. This experiment being repeated when the receiver was inverted in lime-water, and likewife in the infufion of turnfole, there enfued no precipitation, no change of colour, nor any phlogiftication of the air. On pouring ou this air, the ufual fmell of the electric fluid was perceived very fenfibly.

Nitrous air was diminifhed of more than the half of its original bulk, and in that diminifhed ftate, being mixed with common air, it occafioned no red colour, nor any fenfible diminution. It had loft its ufual fmell, and it extinguifhed a candle. In paffing the fparks through the nitrous air, a powder is formed on the furface of the quickfilver, which is a part of that metallic fubftance diffolved by the nitrous acid.

Inflammable air, obtained from iron and
T 4
diluted

diluted vitriolic acid, communicated a little rednefs to the tincture of turnfole. The ftream of electric fluid through this air appeared more red, and much larger, than in common air, being every where furrounded by a faint blue light.

The inflammable air, obtained from fpirit of wine and vitriolic acid, was increafed to about three times its original bulk, and loft a little of its inflammability.

Fixed air, from chalk and vitriolic acid, was a little increafed in bulk by the action of Electricity; but it was rendered lefs abforbible by water.

Vitriolic acid air, obtained from vitriolic acid and charcoal, was diminifhed a little, and black fpots were formed on the infide of the glafs receiver. Afterwards it was obferved, that only one eighth part of the electrified elaftic fluid was abforbed by water. It extinguifhed a candle, and had very little fmell.

Marine acid air feemed to oppofe in great
meafure

meafure the paffage of the electric fluid; fince the fparks would not pafs through a greater length than $2\frac{1}{4}$ inches of this air. It was confiderably diminifhed, but the reft was readily abforbed by water.

Spathous air was neither diminifhed, nor any other way fenfibly altered, by the electric fparks.

Alkaline air, extracted from fpirit of fal ammoniac, was at firft almoft doubled in bulk; then it was diminifhed a little; after which it remained without any augmentation or diminution. It became unabforbible by water, and by the contact of flame it exploded, like a mixture of inflammable air and a good deal of common air.

Common air was laftly tried, and it was found to give a little faint rednefs to the tincture of turnfole; becoming at the fame time fenfibly phlogifticated. The experiment was repeated thrice at different times, and in each time after the electrization it was examined by the admixture of nitrous air in Mr. Fontana's eudiometer, and it

yas

was compared with the fame air not electrified ; the latter always fuffering the
greateft diminution. In the firft experi-
ment the diminutions were $\frac{144}{300}$ and $\frac{175}{300}$; in
the fecond, $\frac{159}{300}$ and $\frac{184}{300}$; and in the laft, $\frac{148}{300}$
and $\frac{171}{300}$.

The experiments laftly defcribed by Mr.
VAN MARUM, were made with a large
battery, which was charged by the above-
mentioned machine. The battery confifted
of 135 coated phials, all together containing about 130 fquare feet of coated furface;
and it was generally charged to the full, by
about 100 turns of the glafs plates of the
machine. This battery appears to have had
a much greater power than what was ever
obtained before by any inftrument of the
kind. With it they melted an iron wire,
15 feet long, and $\frac{1}{151}$ of an inch in diameter ; and another time they melted an iron
wire, 25 feet long, and $\frac{1}{240}$ of an inch in
diameter.

With this extraordinary power of accumulated Electricity, they tried to give magnetifm to needles made out of watch-fprings,
which

which were three and even fix inches in length; and alfo to fteel bars of nine inches in length, from a quarter to half an inch broad, and about a twelfth of an inch thick. Here follows the refult of thofe ingenious experiments.

When the bar or needle was placed horizontally in the magnetic meridian, whichever way the fhock entered, the end of the bar that ftood towards the north acquired the north polarity, or the power of turning towards the north when freely fufpended, and the oppofite end acquired the fouth. If the bar, before it received the fhock, had fome polarity, and was placed with its poles contrary to the ufual direction, then its natural polarity was always diminifhed, and often reverfed, fo that the extremity of it, which in receiving the fhock looked towards the north, became the north pole, &c.

When the bar or needle was ftruck ftanding perpendicularly, its loweft end became the north pole in any cafe, even when the bar had fome magnetifm before, and was placed

with

with the fouth pole downwards. All other circumftances being alike, the bars feemed to acquire an equal degree of magnetic power, whether they were ftruck whilft ftanding horizontally in the magnetic meridian, or perpendicular to the horizon.

When a bar or needle was placed in the magnetic equator, whichever way the fhock entered never gave it any magnetifm; but if the fhock was given through its width, then the needle acquired a confiderable degree of magnetifm, and the end of it which laid towards the weft became the north pole, and the other end the fouth pole.

If a needle or bar, already magnetic, or a real magnet, was ftruck in any direction, its power was always diminifhed. For this experiment, they tried confiderably large bars; one being 7,08 inches long, 0,26 broad, and 0,05 thick.

When the fhock was fo ftrong, in proportion to the fize of the needle, as to render it hot, then the needle generally acquired no magnetifm at all, or very little.

The

The experiments laftly tried with this very powerful battery were concerning the calcination of metallic fubftances, and the revivification of their calces. It appears that the electric fhock produced both thefe apparently contradictory effects.

The metallic calces ufed in thofe experiments were of the pureft fort; they were confined between glaffes, whilft the fhock was paffed over them. By this means the calces were fo far revivified as to exhibit feveral grains of the metal, large enough to be difcerned by the naked eye, and to be eafily feparated from the reft.

As to the calcination of metals, whenever a fhock was employed much greater than that which was neceffary to fufe the metal, part of the metal was calcined, and difperfed into fmoke. It is remarkable, that this calcination or fmoke generally produced feveral filaments, of various lengths and thickneffes, which fwam in the air. It was farther obferved, that thofe flying filaments of metallic calx, if a Conductor was prefented to them, were foon attracted by it; but after the firſt contaƈt they were inſtantly repelled, and generally broke into diverſe parts.

Thus far of this extraordinary machine, to which I ſhall only add a wiſh, that machines ſtill larger may be attempted by the opulent and the ingenious ; for ſince this powerful machine has ſhewn ſeveral particulars both new and intereſting, it is very probable that larger machines and farther experiments will diſcover more of the hidden ſecrets of nature.

N° VII.

N° VII.

OF THE ELECTRIC PROPERTIES OF THE TORPEDO, GYMNOTUS ELECTRICUS, AND SILURUS ELECTRICUS.

THE electric power, by the ancients observed only in amber, and perhaps in the tourmalin, was in procefs of time found to be produced by glafs, fulphur, refins, filk, and diverfe other bodies. It has been within thefe forty years that its great influence has been difcovered in the atmofphere; and not yet fifteen years fince that power has been obferved even in the animal kingdom. Three fifhes have hitherto been difcovered to have the fingular property of giving fhocks analogous to thofe of artificial Electricity, namely, the *torpedo*, the *gymnotus electricus*, and the *filurus electricus*.

The ancients indeed were acquainted with this wonderful property of the firft, and
probably

probably of the laſt of thoſe animals, and it was from the property of giving that ſhock, benumbing ſenſation, or *torpor*, that they called the firſt *torpedo*; but they were utterly ignorant of the cauſe of it. Two diſtinguiſhed writers of the laſt century endeavoured to explain this property upon mechanical laws, but their ingenuity was inſufficient to anſwer for the phenomenon.

The principal diſcoveries, relating to the identity of the above-mentioned property of thoſe fiſhes and the electric ſhock, were made by. JOHN WALSH, Eſq. F. R. S. to whoſe ingenuity we are indebted for the demonſtration of the power of the torpedo agreeing with the electric ſhock in all the points of compariſon in which it could be examined; and alſo for almoſt all the other diſcoveries which were ſince made relating to animal Electricity.

The three animals which, whilſt alive, are poſſeſſed of an electric power, are belonging to three different orders of fiſh; and the few particulars, which they ſeem to have in common, are the power of giving the ſhock;

an

an organ in their body, now called the *electric organ*, which is employed by the animals for the exertion of that power; a fmooth fkin without fcales; and fome fpots here and there on the furface of their bodies.

The torpedo, which belongs to the order of *rays*, is a flàt fifh, very feldom twenty inches in length, weighing not above a few pounds when full grown, and pretty common in various parts of the fea-coaft of Europe. The electric organs of this animal are two in number, and placed one on each fide of the cranium and gills, reaching from thence to the femicircular cartilages of each great fin, and extending longitudinally from the anterior extremity of the animal to the tranfverfe cartilage which divides the thorax from the abdomen. In thofe places they fill up the whole thicknefs of the animal from the lower to the upper furface, and are covered by the common fkin of the body, under which, however, are two thin membranes or *fafciæ*. The length of each organ is fomewhat lefs than one third part of the length of the whole animal. Each organ confifts of perpendicular columns, reaching

Vol. II. U from

from the under to the upper furface of the
body, and varying in length according to the
various thicknefs of the fifh in various parts
The number of thofe columns is not conftant, being not only different in differen
torpedos, but likewife in different ages o
the animal, new ones feeming to be pro
duced as the animal grows. In a very larg
torpedo one electric organ was found t
confift of 1182 columns. The greatei
number of thofe columns are either irregula
hexagons, or irregular pentagons, but thei
figure is far from being conftant. Their di
ameters are generally one fifth part of a
inch. " Their coats are very thin, and feer
tranfparent, clofely connected with eac
other, having a kind of loofe net-work c
tendinous fibres paffing tranfverfely and ob
liquely between the columns, and unitin
them more firmly together : thefe are mof
ly obfervable where the large trunks of th
nerves pafs. The columns are alfo attache
by ftrong inelaftic fibres, paffing directl
from the one to the other."

" Each column is divided by horizonta
partitions, placed over each other at ver
fma.

finall diftances, and forming numerous in-
terftices, which appear to contain a fluid.
Thefe partitions confift of a very thin membrane, confiderably tranfparent. Their
edges appear to be attached to one another,
and the whole is attached by a fine cellular
membrane to the infide of the columns:
They are not totally detached from one another : I have found them adhering at different places, by blood-veffels paffing from
one to another."

" The number of partitions contained
in a column of one inch in length, of a torpedo which had been preferved in proof fpirit, appeared, upon a careful examination; to
be one hundred and fifty : and this number,
in a given length of column, appears to be
common to all fizes in the fame ftate of hu-
midity, for by drying they may be greatly
altered ; whence it appears probable that
the increafe in the length of the column, during the growth of the animal, does not en-
large the diftance between each partition in
proportion to that growth ; but that new
partitions are formed, and added to the extremity of the column from the *fafcia*."

" The

" The partitions are very vafcular; the arteries are branches from the veins of the gills, which convey the blood that has received the influence of refpiration. They pafs along with the nerves to the electric organ, and enter with them; then they ramify in every direction, into innumerable fmall branches upon the fides of the columns, fending in from the circumference all around, upon each partition, fmall arteries, which ramify and anaftomofe upon it; and paffing alfo from one partition to another, anaftomofe with the veffels of the adjacent partitions."

" The veins of the electric organ pafs out clofe to the nerves, and run between the gills, to the auricle of the heart."

" The nerves inferted into each electric organ arife by three very large trunks from the lateral and pofterior part of the brain. The firft of thefe, in its paffage outwards, turns round a cartilage of the *cranium*, and fends a few branches to the firft gill, and to the anterior part of the head, and then paffes into the organ towards its anterior extremity.

mity. The fecond trunk enters the gills between the firft and fecond opening, and, after furnifhing it with fmall branches, paffes into the organ near its middle. The third trunk, after leaving the fkull, divides itfelf into two branches, which pafs to the electric organ through the gills; one between the fecond and third openings, the other between the third and fourth, giving fmall branches to the gill itfelf. The nerves having entered the organs, ramify in every direction, between the columns, and fend in fmall branches upon each partition, where they are loft."

" The magnitude and the number of the nerves beftowed on thefe organs, in proportion to their fize, muft on reflection appear as extraordinary as the phenomena they afford. Nerves are given to parts either for fenfation or for action. Now if we except the more important fenfes of feeing, hearing, fmelling, and tafting, which do not belong to the electric organs, there is no part, even of the moft perfect animal, which, in proportion to its fize, is fo liberally fupplied with nerves; nor do the nerves feem necef-

U 3 fary

fary for any fenfation which can be fuppofed to belong to the electric organs : and, with refpect to action, there is no part of any animal, with which I am acquainted, however ftrong and conftant its natural actions may be, which has fo great a proportion of nerves *."

Thus far with the anatomical defcription of the animal ; we fhall now proceed to defcribe its wonderful electric properties.

The above-defcribed electric organs feem to be the only parts employed to produce the fhock ; the reft of the animal appearing to be only the Conductor of that fhock, as parts adjacent to the electric organs ; and in fact, by artificial Electricity, it has been found that the animal is a Conductor of the electric fluid. The two great lateral fins which bound the electric organs laterally, are the beft Conductors.

If the torpedo, whilft ftanding in water or out of the water, but not infulated, be

* Mr. HUNTER's anatomical obfervations on the torpedo. Phil. Tranf. vol. LXIII.

touched

touched with one hand, it generally communicates a trembling motion or slight
shock to the hand, but this sensation is only
felt in the fingers of that hand. If the torpedo be touched with both hands at the
same time, one hand being applied to its
under, and the other to its upper surface, a
shock in that case will be received, which is
exactly like that occasioned by the Leyden
phial. When the hands touch the fish on
the opposite surfaces, and just over the electric organs, then the shock is the strongest;
but if the hands are placed in other parts of
the opposite surfaces, the shocks are somewhat weaker, and no shock at all is felt
when the hands are both placed upon the
electric organs of the same surface; which
shews that the upper and lower surfaces of
the electric organs are in opposite states of
Electricity, answering to the *plus* and *minus*
sides of a Leyden phial. When the fish is
touched by both hands on the same surface,
and the hands are not placed exactly on the
electric organs, a shock though weak is still
received, but in that case the opposite power
of the other surface of the animal seems to
be conducted over the skin.

The

The ſhock given by the torpedo when in air is about four times as ſtrong as when in water; and when the animal is touched on both ſurfaces by the ſame hand, the thumb being applied to one ſurface, and the middle finger to the oppoſite ſurface, the ſhock is felt much ſtronger than when the circuit is formed by both hands. Sometimes the torpedo gives the ſhocks ſo quickly one after the other, as ſcarcely to elapſe two ſeconds between them; and when, inſtead of a ſtrong determined ſtroke, it communicates only a *torpor*, that ſenſation is juſtly attributed to the ſucceſſive and quick diſcharge of many conſecutive ſmall ſhocks. It is very ſingular, that the torpedo, even when inſulated, ſhould be capable of giving a great many ſhocks to perſons likewiſe inſulated.

This power of the torpedo is conducted by the ſame ſubſtances which conduct Electricity, and is interrupted by the ſame ſubſtances which are not Conductors of Electricity : hence if the animal, inſtead of being touched immediately with the hands, be touched by non-electrics, as wires, two

wet

wet cords, or other Conductors of Electricity, held in the hands of the experimenter, the shock will also be communicated through them. The circuit may also be
formed by several persons joining hands, and
the shock will be felt by them all at the
same time. When the animal is in water,
if the hands are put in the water, a shock
will also be felt, which will be stronger if
one of the hands be brought quite into contact with the fish, whilst the other hand is
kept in the water at a distance from it. In
short, the shock of this animal is conducted
by the same Conductors which conduct
that of a Leyden phial. This shock of the
torpedo may be also conducted by several
circuits at once, and the circuit may be considerably extended; but the shock is always
weakened by lengthening or multiplying the
circuit.

The shock of the torpedo cannot pass
through the least interruption of continuity
in the Conductors which form the circuit;
so that it will not be conducted by a chain,
nor will it pass through the air from one
Conductor to the other, when the distance

between

between them is lefs than the two hundredth part of an inch ; confequently no fpark has ever been obferved to accompany it.

No electric attraction or repulfion could be ever obferved to be produced by the torpedo, nor, indeed, by any of the electric fifhes ; though feveral experiments have been purpofely made with them.

Thefe fhocks of the torpedo feem to depend on the will of the animal ; for each effort is accompanied with a depreffion of his eyes, by which even his attempts to give it to non-conductors may be obferved. It is not known whether both electric organs muft always act together, or only one of them may be occafionally put in action by the will of the animal.

All thefe effects of the torpedo may be imitated by means of a large electric battery weakly charged, and furnifhed with Mr. LANE's electrometer, the balls of which muft be put exceedingly near, or almoft in contact with each other. But the properties of the gymnotus electricus, which we are going to defcribe next, will throw

more

more light on thofe of the torpedo, and will
fhew much more evidently their analogy
with the power of the electric fluid difperfed
through a vaft furface of coated electric fubftance. " But," fays Mr. WALSH, " after
" the difcovery that a large area of rare
" Electricity would imitate the effect of
" the torpedo, it may be inquired, where
" is this large area to be found in the animal?
" we here approach to that veil of nature,
" which man cannot remove. This, how-
" ever, we know, that from infinite divifion
" of parts infinite furface may arife, and
" even our grofs optics tell us, that thofe
" fingular organs, fo often mentioned, con-
" fift, like our electric batteries, of many
" veffels, call them cylinders or hexagonal
" prifms, whofe fuperficies taken together
" furnifh a confiderable area *."

The gymnotus electricus has been frequently called *electrical eel*, on account of its
fuperficial refemblance to the common eel;
though, when accurately examined, it is
found to have none of the fpecific properties
of that animal. The gymnotus is found

* Phil. Tranf. vol. LXIII. p. 476.

pretty frequently in the great rivers of South America. Its ufual length is about three feet; but it has been faid that fome of them have been feen fo large, as to be able to ftrike a man dead with the fhock of their electric organs. A few of thefe animals were brought to England about ten years ago, which, as far as I know, were the firft of the kind brought to Europe. They had been catched in Surinam river, a great way above where the falt-water reaches. It was with thofe identical fifhes that Mr. WALSH made many difcoveries relating to their electrical properties, and the experiments which fhew thofe properties were publicly exhibited in London, during feveral months.

The fubject of Animal Electricity was confiderably advanced by the difcovery of the fpark, with which the fhock of the gymnotus was attended; for, notwithftanding the previous difcoveries relating to the torpedo, and the actual poffibility of imitating the effects of that animal's extraordinary power by means of a large battery weakly charged with artificial Electricity, yet the fcrupulous philofophers ftill fufpected that the power of the torpedo might be fomething different
ent

ent from Electricity, fince the two principal characteriftics of Electricity, namely the
fpark and the attraction, had never been difcovered in the torpedo; and at the fame time
it was difficult to conceive the manner in
which the electric fluid might be generated,
accumulated, and difcharged in an animal,
which, at leaft in its ufual ftate of exiftence,
is a Conductor of Electricity, and is furrounded by a fluid which is likewife a Conductor of that power.   This indeed ftill remains a profound fecret; and it is difficult
to fay, whether any future experiments will
ever difclofe it.   But the fpark having been
difcovered with the gymnotus, the analogy
between its power and Electricity is rendered confiderably more evident, and it
would be fcepticifm to doubt, of the property of the torpedo being derived from the
fame caufe as that of the gymnotus.

In order to proceed regularly, I fhall now
begin with the defcription of the animal,
and fhall then enumerate its electric properties, in a very concife manner.

A gymnotus of three feet length, is generally

rally between ten and fourteen inches in circumference, about the thickeft part of its body. The electric power of this animal being much greater than that of the torpedo, its electric organs are accordingly a great deal larger, and indeed that part of its body which contains moft of the animal parts, or the parts common to the fame order of fifhes, is confiderably fmaller than that which is fubfervient to the electric power; though the latter muft naturally derive nourifhment and action from the former.

The head of the animal is large, broad, flat, fmooth, and impreffed with various fmall holes. The mouth is rather large, but the jaws have no teeth, fo that the animal lives by fuction, or by fwallowing the food entire. The eyes are fmall, flattifh, and of a bluifh colour, placed a little way behind the noftrils. The body is large, thick, and roundifh for a confiderable diftance from the head, and then diminifhes gradually. The whole body, from a few inches below the head, is diftinguifhed into four longitudinal parts, clearly divided from each other by lines. The carina begins a
few

few inches below the head, and widening as it proceeds, reaches as far as the tail, where it is thinneſt. It has two pectoral fins, and the *anus* is ſituated on the under part, more forward than thoſe fins, and of courſe not far diſtant from the *roſtrum*.

This animal has two pair of electric organs, one pair being larger than the other, and occupying moſt of the longitudinal parts of the body. They are divided from each other by peculiar membranes. " The ſtructure of theſe organs is extremely ſimple and regular, conſiſting of two parts, *viz.* flat partitions or *ſepta*, and croſs diviſions between them. The outer edges of theſe *ſepta* appear externally in parallel lines nearly in the direction of the longitudinal axis of the body. Theſe *ſepta* are thin membranes placed nearly parallel to one another: their lengths are nearly in the direction of the long axis, and their breadth is nearly the ſemidiameter of the body of the animal. They are of different lengths, ſome being as long as the whole organ. I ſhall deſcribe them as beginning principally at the anterior end of the organ, although a few begin

5   along

along the upper edge; and the whole, paffing towards the tail, gradually terminate on the lower furface of the organ, the lowermoft at their origin terminating fooneft. Their breadths differ in different parts of the organ. They are in general broadeft near the anterior end, anfwering to the thickeft part of the organ, and become gradually narrower towards the tail; however, they are very narrow at their beginnings or anterior ends. Thofe neareft to the mufcles of the back are the broadeft, owing to their curved or oblique fituations upon thefe mufcles, and grow gradually narrower towards the lower parts; which is in a great meafure owing to their becoming more tranfverfe, and alfo to the organ becoming thinner at that place. They have an outer and an inner edge: the outer is attached to the fkin of the animal, to the lateral mufcles of the fin, and to the membrane which divides the great organ from the fmall; and the whole of their inner edges are fixed to the middle partition, to the air bladder, and three or four terminate on that furface which inclofes the mufcles of the back. Thefe *fepta* are at the greateft diftance from one another at their exterior

2   edges

edges, near the fkin, to which they are united; and as they pafs from the fkin towards
their inner attachments they approach one
another: fometimes we find two uniting
in one. On that fide next to the mufcles
of the back they are hollow from edge to
edge, anfwering to the fhape of thofe mufcles, but become lefs and lefs fo towards
the middle of the organ; and from that towards the lower part of the organ they become curved in the other direction. At
the anterior part of the large organ, where it
is nearly of an equal breadth, they run pretty parallel to one another, and alfo pretty
ftraight; but where the organ becomes narrower, it may be obferved in fome places
that two join or unite into one, efpecially
where a nerve paffes acrofs. The termination of this organ at the tail is fo very fmall,
that I could not determine whether it confifted of one *feptum* or more. The diftances between thefe *fepta* will differ in fifhes of different fizes. In a fifh of two feet four
inches in length I found them $\frac{1}{27}$ of an inch
diftant from one another, and the breadth of
the whole organ, at the broadeft part, about
an inch and a quarter, in which fpace were

34 *septa*. The small organ has the same kind of *septa*, in length passing from end to end of the organ, and in breadth passing quite across, they run somewhat serpentine, not exactly in straight lines. Their outer edges terminate on the outer surface of the organ, which is in contact with the inner surface of the external muscle of the fin, and their inner edges are in contact with the centre muscles. They differ very much in breadth from one another; the broadest being equal to one side of the triangle, and the narrowest scarcely broader than the point or edge. They are pretty nearly at equal distances from one another; but much nearer than those of the large organ, being only bout $\frac{1}{56}$ part of an inch asunder: but they are at a greater distance from one another towards the tail, in proportion to the increase of breadth of the organ. The organ is about half an inch in breadth, and has fourteen *septa*: these *septa*, in both organs, are very tender in consistence, being easily torn; they appear to answer the same purpose with the columns in the *torpedo*, making walls or butments for the subdivisions, and are to be considered as making so many dis-

tinct

tinct organs. Thefe *septa* are interfected tranfverfely by very thin plates or membranes, whofe breadth is the diftance between any two *septa*, and therefore of different breadths in different parts; broadeft at that edge which is next to the fkin, narroweft at that next to the centre of the body, or to the middle partition which divides the two organs from one another. Their lengths are equal to the breadths of the *septa* between which they are fituated : there is a regular feries of them continued from one end of any two *septa* to the other; they appear to be fo clofe as even to touch. In an inch in length there are about 240, which multiplies the furface in the whole to a vaft extent *."

The nerves which go to the electric organs of the gymnotus, as well as of the torpedo, are much larger than thofe which fupply any other part of the body. The electric organs of the gymnotus are fupplied with nerves from the fpinal marrow, and

* Mr. HUNTER's account of the Gymnotus Electricus. Phil. Tranf. Vol. LXV.

they

they come out in pairs between the vertebræ of the fpine.

The gymnotus poffeffes all the electric properties of the torpedo, but in a fuperior degree. His fhock is conducted by the Conductors of Electricity, and interrupted by the non-conductors of the electric fluid. Hence the fhock is communicated through water, without the immediate contact of the animal, or through any other proper circuit; but the ftronger fhock is received by touching the animal when out of the water; and the beft way to receive ftrong fhocks, is to apply one hand towards the tail, and the other towards the head of the animal. In this manner I have often received fhocks, which I felt not only in my arms, but even very forcibly in my breaft. If the animal be touched only with one hand, then a kind of tremor is felt only in that hand, which though ftronger, is quite analogous to the fenfation communicated by the torpedo when touched in the like manner. The gymnotus's power of giving fhocks is alfo depending on the will of the animal, fo that fometimes he gives very ftrong fhocks,

and

and at other times very weak ones, but he gives the ftrongeft fhocks when provoked by being frequently and roughly touched.

When fmall fifhes are put into the water wherein the gymnotus is kept, they are generally ftunned or killed by a fhock, and then they are fwallowed, if the animal is hungry. The fifhes which are ftunned by the gymnotus may often be recovered by being fpeedily removed into another veffel of water.

The ftrongeft fhocks of the gymnotus will pafs a very fhort interruption of continuity in the circuit. Thus they will be conducted by a chain, efpecially when it is not very long, and is ftretched, fo as to bring its links into better contact. When the interruption is formed by the incifion made by a pen-knife on a flip of tin-foil that is pafted on glafs, and that flip is put into the circuit, the fhock in paffing through that interruption will fhew a fmall but vivid fpark, plainly diftinguifhable in a dark room.

X 3   Mr.

Mr. WALSH made another remarkable
difcovery with the gymnotus, which he
fhewed at his houfe to various ingenious perfons : it was a new fort of fenfe in the animal, by which he knew when the bodies
which came near him were fuch as could
receive the fhock, (*viz.* Conductors) and
when they were of the contrary nature; in
the former of which cafes the animal gave
the fhock, but not in the latter. In order
to fhew this wonderful property, divers experiments were made, but the moft convincing one was the following :—the extremities
of two wires were dipped into the water of the
veffel wherein the animal was kept, then they
were bent, and extended a great way, and
laftly terminated in two feparate glaffes full of
water. Thefe wires being fupported by
non-conductors at a confiderable diftance
from each other, it is plain that the circuit
was not complete ; but if a perfon put the
fingers of both his hands into the glaffes
wherein the wires terminated, *viz.* thofe of
one hand into one, and thofe of the other
hand into the other glafs, then the circuit
was complete. Now it was conftantly obferved, that whilft the above-defcribed circuit

cuit remained interrupted, the animal never went purpofely near the extremities of the wires, as he ufed to do when willing to give the fhock : but the moment that the circuit was completed, either by a perfon or any other Conductor, the animal immediately went towards the wires, and gave the fhock; though the completion of the circuit was performed quite out of his fight.

Several other particulars, not only concurring to prove the above-mentioned property, but otherwife interefting, were afcertained by Mr. WALSH; but for thefe we muft wait till that ingenious gentleman will favour the public with a particular account of his refearches.

The third fifh which is known to have the power of giving the fhock, is found in the rivers of Africa, but we have a very imperfect account of its properties *.

* Meffrs. ADANSON and FORSKAL make a fhort mention of it, and Mr. BROUSSONET defcribes it, under the French name of *le Trembleur*, in the Hift. de l'Academie Royale des Sciences, for the year 1782.

This

This animal belongs to the order which the naturalifts call *filurus*; hence it is commonly called *filurus electricus*. Some of thofe fifhes have been feen even above twenty inches long.

The body of the filurus electricus is oblong, fmooth, and without fcales; being rather large, and flattened towards its anterior part. The eyes are of a middle fize, and are covered by the fkin, which envelopes the whole head. Each jaw is armed with a great number of fmall teeth. About the mouth it has fix filamentous appendices, *viz.* four from the under lip, and two from the upper; the two external ones, or farthermoft from the mouth on the under lip, are the longeft. The colour of the body is greyifh, and towards the tail it has fome blackifh fpots.

The electric organ feems to be towards the tail, where the fkin is thicker than on the reft of the body, and a whitifh fibrous fubftance, which is probably the electric organ, has been diftinguifhed under it.

3   It

It is faid that the filurus electricus has the property of giving a fhock or benumbing fenfation like the torpedo, and that this fhock is communicated through fubftances that are conductors of Electricity; but no other particular about it is known with any confiderable degree of certainty.

An inquifitive mind will immediately enquire, for what purpofe has nature furnifhed thofe animals with fo fingular a property. But the prefent knowledge of the fubject feems to furnifh no other anfwer, except, that they are endowed with the power of giving the fhock for the fake of fecuring their prey, by which they muft fubfift, and perhaps of repelling larger animals, which might otherwife annoy them.

The ancients confidered the fhocks given by the torpedo as capable of curing various diforders; and a modern philofopher will hardly hefitate to believe their affertions, after that Electricity has been found to be a remedy for many difeafes.

THE

THE

INDEX.

A..

ABSCESSES electrified, II. 153.

Agues electrified, II. 155.

Air, generally electrified, 72 ; is displaced and rarefied
by a shock, 274; to charge a plate of, 285; when
hot becomes a conductor, 324; how electrified, 326;
effects produced by sparks in it, II. 281.

Air-pump, used for electrical experiments, 237, 238,
241 ; the patent one recommended, 243.

Alkaline air, effects produced by the electric sparks in
it, II. 281.

Amalgam for the excitation of glass, 142.

Animal electricity, II. 287.

Animals electrified, 43, 44, 47.

Apparatus, electrical, in general, 134.

Atmosphere, electrical, 209, 212, 215, 217 ; rendered
visible, 241.

Atmospherical electricity, 71, II. 37 ; electrometer, II.
40, 42, 269.

Attraction, electrical, 2, 16; between bodies possessed of
different electricities, 43, 195, 199 ; explained, 109 ;
shewn by experiments, 190, 333; in vacuo, 196. II.
92, 95 ; exerted to a great distance, II. 275; not
observable with the electric fishes, II. 298.

Aurora borealis, supposed to be an electrical phenomenon,
75; when strong it has been known to have disturbed
the magnetic needle, 75 ; imitated, 238 ; does not
seem to influence the electricity of an electrical kite,
II. 38.

B. *Baked*

I N D E X.

INDEX.

INDEX.

marked with prifmatic colours by the electric fhock,
63, 267; when thin is apt to be broken by a difcharge,
147; a fort of, which will not hold a charge, 149;
globes and cylinders have fometimes burft in whirl-
ing, 138; pillars ought to be varnifhed, 180; metals
ftruck into it by means of a fhock, 279; ftained by
the fhock, 280.

Gold leaf, melted by a fpark from a prime conductor, II.
277.

Gout electrified, II. 155.

Gun-powder fired by means of Electricity, 277. II. 241,
242, 276.

Gutta ferena electrified, II. 148.

Gymnotus electricus, its properties, II. 287.
   defcribed, II. 299, 301.

H.

Hail often electrified, 74.

Hair, its electricity, II. 65.

Head-aches electrified, II. 154.

Heat pervades fome bodies eafier than others, 117.

Heating of certain fubftances produces Electricity, 24,
29, 32.

Hypothefis of Electricity, 106.

I.

Jar, electric, what, 56; it holds a greater charge in condenfed air, 57; a large one is better than many fmall
ones, 148; one particularly defcribed, 175; when hot
it holds a better charge, 186; caution in difcharging
them, 186, 188; often broken by the difcharge, 186;
remedy to prevent their breaking, 186; method of re-
pairing them when cracked, 187; are more eafily
broken when any cement is upon them, 188; when
infulated cannot be charged, 248; charged with its
own electric fluid, 250; in what manner feveral may
be charged almoft as eafily as a fingle one, 252; the
air contained in it is not expelled by the charge, 254;
its

I N D E X.

I. ⋅   Y   *Light,*

V. *Vacuum*,

INDEX.

V.

Vacuum, its properties respecting Electricity, 11, 12; Leyden, 261; experiment made in it by the author, II. 90.

Vapour of hot water is a conductor, 321.

Varnish is not proper for the purpose of coating the inside of electric jars, II. 68.

Vegetable life, in what manner is affected by Electricity, 45.

Velocity supposed to be acquired by the electric fluid in going through long Conductors, II. 232.

Venereal disease electrified, II. 158.

Vessels, in what manner may be secured against the lightning, 83.

Vindicating Electricity, 360; negative, what, II. 197; positive, what, II. 198.

Vitreous Electricity, what, 19.

Vitriolic-acid-air, effects produced by the electric sparks on it, II. 280.

Ulcers electrified, II. 149.

W.

Water, sparks may be rendered visible in it, 272; how affected by a shock, 62, 272, 273, 288; spout imitated, 288; concussion given to it by a shock, 320; illuminated by a shock, 336.

Wax excited by melting, 26; is apt to attract the ashes when melted by the wax-chandlers, 129; sealing, see *Sealing-wax*.

Well, electric, what, and its properties, 201.

Whirlwind, supposed to be an electrical phenomenon, 75; imitated, 289.

Wires, metallic, melted by Electricity, 310. II. 282; may be lengthened and shortened, by means of Electricity, 312; particular sort of conducting ones useful in Medical Electricity, II. 130.

ADDI-

ADDITIONS.

THE mineral acids are much better conductors than other liquids, and the conducting power of water is confiderably increafed by the admixture even of a fmall quantity of any of them.

Quickfilver excites glafs negatively.

Very good tourmalins, it is faid, have been found in the mountains of Saxony, near Freyberg.

To Chap. V. Exp. I.—The pin or pointed body that is prefented to the rubber muft not be very fharp; otherwife the pencil of light is rarely obferved.

To Vol. II. p. 82.—The lower extremity of the filver wire muft be formed into a globule, in order to prevent the corks getting off. This is eafily done, by putting the extremity of the wire in the flame of a candle.

	ERRATA.	CORRIGE.
	Page. Line.	
Vol. I.	75. 11. accentions	accenfions.
	110. 25. wil	will.
	141. 2. radiufes	radii.
	173. 8. flided	flid.
Vol. II.	279. 12. ou	out.

DIRECTIONS to the BINDER.

Take care to place the Plates I. and II. at the end of Vol. I. and the Plates III. IV. and V. at the end of Vol. II.

www.ingramcontent.com/pod-product-compliance
Lightning Source LLC
Chambersburg PA
CBHW021124270326
41929CB00009B/1028